Directions in Human/Computer Interaction

HUMAN/COMPUTER INTERACTION
A Series of Monographs, Edited Volumes, and Texts

SERIES EDITOR
BEN SHNEIDERMAN

Directions In Human/Computer Interaction
Edited by Albert Badre and Ben Sheiderman

Online Communities:
A Case Study of the Office of the Future
Starr Roxanne Hiltz

Human Factors In Computer Systems
John Thomas and Michael Schneider

DIRECTIONS IN HUMAN/COMPUTER INTERACTION

EDITED BY

ALBERT BADRE
School of Information and Computer Science
Georgia Institute of Technology

BEN SHNEIDERMAN
Department of Computer Science
University of Maryland

ABLEX PUBLISHING CORPORATION
Norwood, New Jersey 07648

Library of Congess Cataloging in Publication Data
Main entry under title:

Directions in human/computer interaction.

  (Human/computer interaction; v. 1)
  Includes index.
  1. Interactive computer systems—Addresses, essays,
lectures. 2. System design—Addrsses, essays, lectures.
I. Badre, Albert. II. Shneiderman, Ben. III. Series.
QA76.9.I58D57   1981   001.64   82-11575
ISBN 0-89391-144-5

Printed in the United States of America.

ABLEX Publishing Corporation
355 Chestnut Street
Norwood, New Jersey 07648

Contents

Preface

A. BADRE
B. SHNEIDERMAN

The papers in this book were inspired by a workshop on human-computer interaction that was organized by Albert Badre and supported by a contract award from the U.S. Army Research Institute for the Behavioral and Social Sciences to the Georgia Institute of Technology. This workshop was held in Atlanta in March of 1981.

The purpose of the workshop was to review, consider, and exchange ideas on current research in the general area of interactive system design and human information processing. We attracted forty research scientists from the fields of computer science and psychology of whom twenty reported their results on current research.

The theme of the workshop derived in part from the awareness that there is a growing shift in who uses computers from the computer-trained individual to the person who has no data processing or computer experience. Users of home or personal computers, office automation equipment, and information resources are very different from professional programmers. Yet, the language that the operator must use to interact with the machine, the documentation that he/she has to read and the system messages printed at a terminal are couched in the vocabulary and language habits of the computer expert.

A second motivation for increased concern about human factors issues is the increased dependence of many organizations on their computer and information systems. Higher demands for system reliability and data integrity are propelling many organizations to focus on the human aspects of command languages, data entry, response time, system displays and computer program comprehensibility. Airline, hotel, and car reservation systems, banking applications, insurance, inventory, purchasing, payroll, or decision support applications which are increasingly done online, must be designed to ensure higher operator accuracy, speed and satisfaction.

The third motivation for a human factors emphasis is the use of interactive systems in life critical applications such as medical intensive care units, nuclear, air traffic, or utility control, fire or police dispatch, and manned spaceflight. In these environments action must be rapid, commands must be correct, and operators cannot consult lengthy training guides when emergency conditions arise.

There is a growing consensus in the computer science community that human factors considerations should be incorporated into the design of all computer systems at the initial stages of development. ''Information processing'' systems are likely to be more user compatible if they are designed to adapt to the information processing capabilities and limitations of the user. It is becoming, therefore, increasingly necessary to explore and identify the human information processing skills, constraints, and variables that are associated with making the interface more user compatible. This means identifying factors relating to what the operator does at the display station to perform a desired task and how the system responds.

For most current systems, the mode of communication is predominantly textual. For example, an operator may interact with an operating system by typing command statements or with a database by stating queries in a language that can be parsed by the machine. The machine, on the other hand, responds to the operator by displaying on the screen written messages, words, codes, sentences, and passages.

These processes are effective, but increasingly graphics systems, speech input/output, and other novel ideas are being explored for use in interactive applications. In this book the authors focus on improving contemporary approaches as well as exploring emerging design strategies.

—Robert B. Allen begins with a broad ranging review of research and numerous references which bridge the gap between computer science and cognitive psychology.

—Elliot Soloway, Kate Ehrlich, Jeffrey Bonar, and Judith Greenspan describe an experiment in Pascal programming and offer deep insights to the difficulty of composing programs.

—Ben Shneiderman offers guidelines for the design of system messages and the results of five small experiments in which COBOL diagnostic, text editor, or job control language messages were altered.

—Paul R. Michaelis, Mark L. Miller, and James A. Hendler argue for a stronger linkage between artificial intelligence and human factors engineering, for applications such as natural language interaction and tutoring.

—Susan T. Dumais and Thomas K. Landauer investigate the naturalness of terminology for command languages and database information retrieval.

—Sam Ehrenreich and Theodora A. Porcu present an important experiment on abbreviation strategies for command languages. They studied the effectiveness of truncation and contraction, fixed and variable lengths, and an approach to handling suffixes.

—Michael L. Schneider theorizes about the process of acquiring skill with a command language and draws conclusions which are relevant for designers.

—Paul R. Michaelis and Richard H. Wiggins provide a lucid and thorough introduction to speech synthesis.

—Albert Badre reports on two experiments which reveal human chunking strategies in a sequence of graphic displays. His results are useful to designers of complex graphical displays.

—Beverly G. Knapp, Frank L. Moses, and Leon H. Gellman review issues and present design recommendations for formatting complex graphic displays.

These papers are a lively mixture of survey, theory, practical recommendations, and experimental results. We believe that they reflect the diversity of research efforts in human computer interaction. This field is rapidly expanding and there is room for many approaches. We hope that these examples stimulate further work and greater sensitivity to the problems of designing interactive computer systems.

We are grateful to our colleagues who volunteered to review and comment on the papers in this book: Nancy Anderson, George Cherry, Roger Ehrich, Jim Foley, Rex Hartson, William S. Mosteller, Carol Anne Ogdin, G. Michael Schneider, Bonnie Webber, Sherry Weinberg, Mark Weiser. These comments, as well as reviews by the authors, strongly influenced the revisions. Especial thanks go to Bonnie Webber who read and provided valuable comments on every paper. Robert Allen deserves credit for providing the technical skills and computer resources to help get several of the computer files into the proper format for the typesetter. The staff of Ablex were always helpful, prompt, and competent.

Directions in Human/Computer Interaction

Cognitive Factors in Human Interaction with Computers

ROBERT B. ALLEN

Bell Laboratories,
Murray Hill, NJ 07974

Designing computer interfaces to match human cognitive processes is increasingly important as computer systems become more sophisticated. This paper examines experimental results, models, and research strategies relevant to cognitive processes in user interfaces for topics including query languages, command languages, programming, problem solving, editing, and displays.

1. INTRODUCTION

The study of human factors has been intertwined with computing science for many years; however, recently new areas of research on human-computer interaction have begun to emerge. These new fields are related to human cognitive activities such as language, problem solving, memory, and attention. Indeed many recent studies on the human factors of computing systems have included dependent variables traditionally associated with cognitive psychology such as human response time, the nature of errors, the time to learn, and self-reports. Moreover, many of the findings and hypotheses from human-interface research are explained in terms of cognitive theories and principles.

Among the topics with cognitive implications that have been studied extensively are programming techniques, command languages, data-base access, and editing. There is a cognitive component even in simple keying tasks (Conrad, 1966; Zeigler & Sheridan, 1965). However, cognitive factors are probably more important for complex tasks, and increasingly software is being applied to these.

1.1 Informal Prescriptions for Human Interfaces

Practitioners of the art of constructing human interfaces have proposed quasi-cognitive criteria for the effective design of systems. Factors such as simplicity, naturalness, ease of use, consistency, feedback, and individualization are frequently and perhaps rightfully mentioned. However, there are potential problems with these informal prescriptions; the details for applying them are rarely specified and there is little independent validation of the assertions. Clearly, a systematic approach to user interfaces would be fruitful; thus, this paper will focus on empirical studies of cognitive processes in human interaction with computers.

1.2 Cognitive Theories

A general review of cognitive theories may be obtained from sources such as Lindsay and Norman (1977) or Monsell (1981). For the purposes of this paper, it will be especially helpful to consider current theories of short-term memory, long-term memory, and problem solving. These areas have been cited particularly frequently, and are related to a wide range of issues in human interaction with computers.

1.2.1 Short-Term Memory. The concept of short-term memory applies to recalling information soon after it has been presented. For instance, remembering displayed command options for a few seconds could involve short-term memory. The short-term memory model proposed by Miller (1956) has become particularly well known in the human-interface literature, but there are several limitations to that model. Miller proposed that human short-term memory resides in a discrete buffer that contains 7 ± 2 items. However, there is much experimental evidence that does not support this. For example, the number of items it is possible to recall decreases as the complexity of the items increases, thus it is possible to remember seven digits, but only three or four disconnected sentences (Simon, 1974). For discussions of other problems with the single-buffer model, see Broadbent (1975) and Monsell (1981). While theorists continue to agree that the capacity of short-term memory is limited, the discrete-buffer models of short-term memory have been replaced by models in which memory resources are dynamically allocated (e.g., Moray, 1967). According to these models, performance may be affected by the ways in which active processes share resources (Norman & Bobrow, 1975). While these resource-allocation models do fit the experimental data more readily, their predictive power is low and, indeed, some of their implications also have been questioned (Duncan, 1980; Moray, 1977).

1.2.2 Long-Term Memory. One pervasive finding in long-term memory research is the effect of the organization of information; naturally this has also been an important characteristic of many models of long-term memory. For example, the process of chunking, mentioned in Miller's (1956) paper, is one type of memory organization process. However, many other memory organization processes have been proposed. One class of models employs associative networks (e.g., Collins, 1975; Tulving, 1972). Another type of model employs schemas (also called frames, prototypes, and scripts) which are memory structures that hold sets of information. Because schemas are becoming increasingly common in the literature on human interaction with computers, the remainder of this section will examine them.

Many specific schema models have been proposed, however they differ in the areas to which they are applied and in the degree of psychological validity attributed to them (Abelson, 1981). For instance, the scripts of Schank and Abelson (1977) relate to the complex, real-world scenarios, while the frames of Minsky (1975) relate to relationship of visually perceived objects. Although the heuristic value of schemas for organizing complex material is obvious, the processes associated with schemas (their use and creation) are actively debated (Chafe, 1975; Feldman, 1975).

As the field of human interaction with computers becomes more concerned with the mental organization of complex material, schema models are becoming increasingly popular. Variations of these models have been applied to command languages (Thomas & Carroll, 1981), to programming (Atwood & Ramsey, 1978), to learning to program (Mayer, 1976, 1981), and to learning the use of text editors (Bott, 1979). Several of these schema models are discussed in more detail below.

1.2.3 Problem Solving. Another approach to complex cognitive processes focuses on the activities associated with problem solving. Although there are several approaches to the cognitive processes involved in problem solving (e.g., Greeno, 1973; Thomas & Carroll, 1979), the production-system models of Newell and Simon (e.g., 1972) are the most widely known. Production-system models assert that when a given state (condition) is achieved in short-term memory, a certain production (action) will be triggered.

The early production-system models were applied to highly constrained tasks such as cryptarithmetic, and the ''Towers of Hanoi'' problem (Hayes & Simon, 1974). However, recent work has applied variations of the the models beyond problem solving to many types of complex cognitive activities including several that involve human interaction with computers. Among these are models to describe programming (Brooks, 1977), editing (Card, Moran, & Newell, 1980a), reading (Just & Carpenter, 1980), speech perception (Newell, 1980), text composition (Hayes & Flower, 1980), and fault diagnosis (Rouse, Rouse, & Pellegrino, 1980). Several of these recent efforts will be discussed below.

Because production-system models involve complex cognitive processes, they are often based on verbal protocols that are descriptions by subjects of what they are thinking. Obviously, there are several levels at which production-system models may be evaluated; they may be examined in terms of the net sufficiency of the model or they may be examined in terms of the details of the cognitive processes involved. The extent of their validity as psychological models has been the object of extensive debate (e.g., Davis & King, 1977; Newell, 1980; Norman,

1980; Young, 1979). One troublesome area for formal production-system models concerns the capacity of short-term memory. As noted above, models of short-term memory with discrete buffers generally have been replaced by resource-allocation models. Moreover, the production-system models themselves are not consistent about the capacity of short-term memory; for example, Brooks' model requires that short-term memory holds 14 items.

In Section 2, specific studies involving cognitive factors in human interaction with computers will be examined. In Section 3, some of the limitations and implications of the approach will be discussed.

2. EMPIRICAL STUDIES

2.1 Directing a Computer's Actions

2.1.1 Data-Base Access Languages. Human-interface issues in data-base search languages have been extensively studied. This was a natural area for early human factors research because self-contained data-base search programs can be employed by novice computer users. Rather than attempting to review the entire literature here (see Reisner, 1981; Shneiderman, 1980a), those issues most relevant to cognitive processes will be emphasized.

In a study of SQUARE and SEQUEL, Reisner (1977) reports that subjects frequently made errors by attempting to derive the syntax of their queries from English syntax. In addition, subjects often incorrectly generalized principles of the data-base language from one application to another. As a framework for describing these errors, Reisner (1977, 1981) proposes a model based on the work of Chomsky (1964), similar in some respects to the schema models of long-term memory. The model asserts that a user generates a mental template of the structure of a query and then proceeds to determine specific parameters to be inserted in the template. Errors arise from the difficulty of transforming natural English questions into the format required by the query language.

Apparently many errors and the time to learn a query language can be reduced by providing an explicit structure for the user to complete, as in Query-by-Example (Thomas & Gould, 1975). Among the advantages of this type of input are that the user is prompted about the types of information required, and in most cases the syntax is easy to recall.

For Query-by-Example (Thomas, 1976) as well as for many other data-base systems (Gould & Ascher, 1975; Reisner, 1977; Thomas & Gould, 1975), a recurring problem involves quantification and set membership. Few users are familiar with the formal-logic concepts implicit in many data-base search systems. Indeed for many types of tasks, natural logic has been found to be greatly different from formal logic (Braine, 1978). Sometimes it may be possible to tailor query languages to match an individual's problem-solving strategy; however, even this must be considered a limited solution since many individuals are not consistent in their natural logic (Thomas, 1976).

2.1.2 Command Languages.

Several recent studies have focused on command languages; a major issue in these is the degree to which command languages should resemble natural language in terminology and structure (Treu, 1975). Ledgard, Whiteside, Singer, and Seymour (1980) compared the use of an editor with a compact notation to an editor with English-like terminology. With the English-like command language, subjects completed more editing tasks and made fewer errors than with the other editor, even though the English-like editor required more keystrokes. On the other hand, Landauer, Galotti, and Hartwell (1980) did not find significant differences in initial learning and performance between text editors employing either familiar (frequently selected) command terms or arbitrarily selected terms.

Several studies have suggested the importance of consistency and rules in command languages. For instance, Streeter, Ackroff, and Taylor (1980) report that command abbreviations generated by rule are more easily recalled than commonly used command abbreviations. In another study of command abbreviations, Ehrenreich and Porcu (1982) demonstrate that truncation of command terms leads to both better encoding and decoding than contraction of command terms. Likewise, fixed-length abbreviations were found to be better than variable-length abbreviations. Thomas and Carroll (1981) propose that individuals use schemas, which may be considered a type of rule, for command terms. For example, they suggest that "congruence" is an important property for a command language; by this they mean that commands should be paired with their semantic opposites when opposite actions are available (e.g., "open" and "close"). Barnard, Hammond, Morton, Long, and Clark (1981) examined the importance of word order in commands with two arguments. Overall, they found that consistency in the position of a recurrent argument led to better performance, at least when the recurrent argument was first.

Many of the results from experiments on command languages may be understood in terms of reconstructive memory. While an experienced computer user may be able to retrieve the correct command from memory, users who can't recall the command perfectly may make an inference based on what they do remember and their expectations about command terms. Thus, if the user knows simple rules for generating commands these will lead to correct inferences. Barnard et al. (1981) and Bott (1979) provide additional discussions of reconstruction processes for command terms.

Beyond the reconstruction of command terms, there are several other types of cognitive processes involved in command-language use. For instance, the user must analyze the task, a particular command must be selected (see the GOMS Models discussed in Section 2.4.2), and the command must be entered to the computer (see the Keystroke-Level Model discussed in Section 2.4.2). Although most previous studies have focused on one or another of these processes, it is useful to realize that each of them may affect command-language performance. For instance, studies of errors in command languages have tended to lump all errors together, but different types of errors can be traced to the different cognitive activities. Likewise, experience with computer systems may be examined in terms of the different types of cognitive activities involved. Finally, there may be interactions among the different cognitive activities; for instance, high processing demands for the reconstruction of a command term may withdraw attentional resources from the entry of the commands and lead to more errors.

2.1.3 Menus and Trees. Menus are simply lists of options available to a computer user. Menus may be used in interactive control tasks (e.g., Gade, Fields, Maisano, Marshall, & Alderman, 1981), or alternatively, they may provide a data-base search interface by having categories of information arranged into networks (McCracken & Robertson, 1979) or trees.

Menus and trees incorporate several desirable psychological features (Allen, 1981a; Grimes, 1979), such as allowing users simply to recognize items rather than to recall them. Studies of simulated retrieval systems have shown that hierarchical arrangements of information are particularly appropriate for certain types of material (Barnard, 1978; Brosey & Shneiderman, 1978; Durding, Becker, & Gould, 1977) and for certain individuals (Broadbent & Broadbent, 1978). In a study of the use of computerized menu and tree system, Allen (1981a) found patterns of response times that suggested cognitive processing. For example, there was a longer pause prior to long searches than short searches, thus subjects may have been planning the searches during that time.

2.1.4 Natural-Language Interfaces. There are many obstacles in the development of natural-language interfaces. For the foreseeable future, natural-language systems will be able to deal with only subsets of natural language. There are important human-factors questions, then, in the way people can adapt to these restricted subsets (Watt, 1968).

Of course there are different aspects of a language that may be restricted; one approach is to allow only a limited vocabulary. For participants in a problem-solving task, Kelley and Chapanis (1977) determined that a task-relevant vocabulary of 300 words was as effective as an unrestricted vocabulary. Malhotra (1975) recorded subjects' questions while using an unrestricted vocabulary to a simulated management-information system. He found that in 496 queries, 358 unique words were used. Extrapolating the rate at which unique words were introduced, Malhotra estimated that a vocabulary of 1,000 to 1,500 words should be sufficient for most queries to this data base. Thus, for many applications a relatively small vocabulary is sufficient and when necessary, users can restrict their vocabularies.

Observational studies also suggest that it should be possible to restrict the syntactic constructions of queries. Malhotra (1975) found that 78% of the queries posed to his management-information system were of 10 basic syntactic types; indeed 55% of all sentences were of only two types. In a study of how users referred to or defined words, Dumais and Landauer (1982) report that people rarely use negatives or opposites in their definitions; rather, the definitions typically specify a superordinate that subsumes the word, and a few examples or attributes of the word.

In addition to vocabulary and syntax, a natural-language system must have a data base, models for the relationship among items in the data base, and models for the requirements of the user (Winograd, 1980). While many elaborate models for such world knowledge have been developed (e.g., Schank & Abelson, 1977), few studies have considered the effects on usability of variations in these models (Malhotra, 1975).

A few experimental studies have directly compared natural language to other types of input. Small and Weldon (1977) gave subjects data-retrieval tasks to be completed with either a restricted natural English or the data-base language SEQUEL. The subjects were equally accurate with the two techniques but were somewhat faster using SEQUEL for the tasks in which the task was highly structured. While this experiment provides a bench mark for further research and it is surprising that subjects using the data-base language did better than with English, conclusions must be drawn cautiously. For example, the superiority of SEQUEL appeared only in a group that had previously used English input.

Beyond these empirical studies the possibility of communicating with computers in a fashion that resembles natural human-to-human language has been the object of hope (Petrick, 1976) and criticism (Fitter, 1979; Shneiderman, 1980b). Proponents of natural-language interfaces emphasize the range of possible users that could interact with computers without extensive training. Critics of natural-language interfaces point out that concise, rather than natural, languages may be more efficient for many types of problem solving and, of course, natural-language interfaces require substantial processing-time and development costs.

2.1.5 *Interaction with Computers as Conversation.* Another approach to natural-language interfaces proposes that dialogue with a computer may be compared to conversations among people (e.g., Ballantine, 1980; Nickerson, 1976). Several authors have suggested that it is more important to extract the intention and goals of the interaction than to extract the literal meaning (Thomas, 1978; Winograd, 1980). One framework for analyzing intention in conversations employs ''speech acts'' (Searle, 1969) and may be applied to natural-language interfaces (Winograd, 1980). An understanding of the dynamics of conversations may also improve the ability to model the context of language use. Grosz (1978) proposes that a conversation proceeds by channeling the speaker's attention in a proc-

ess she calls "focusing." In a similar vein, Thomas (1978) proposes that the structure of a conversation is directed by explicit "metacomments."

2.2 Cognitive Factors in Programming

2.2.1 Cognitive Models of Programming. In their efforts to improve the efficiency of programmers, software theorists often have invoked cognitive principles. For example, the concept that the capacity of short-term memory is constant has been used to support innovations such as structured programming and right-branching program structures (e.g., Tracz, 1979). Short-term memory limitations clearly can be an important factor in programming. For example, Sime, Green, and Guest (1973) found that nested conditional statements (if-then-else) reduced error rate and solution time as compared to branch-to-label if statements. They explain this result on the basis of lower short-term memory demands for the nested statements.

Because there is such a large literature on programming, this section will focus on studies that are related to cognitive models. However, the interested reader may refer to general reviews (Sheil, 1981; Shneiderman, 1980a), to the discussions of specific cognitive effects in programming (Green, 1980; Shneiderman & Mayer, 1979), and to the methodological critique of Brooks (1980).

Atwood and Ramsey (1978) apply Kintsch's (1978) schema theory of text comprehension to program comprehension. Thus, the Atwood and Ramsey model suggests that subjects formulate a hierarchical schema of program statements. An experimental test of the model was conducted by Atwood and Ramsey. This test examined various explanations for differences in debugging times for different types of program bugs (Gould & Drongowski, 1974). Unfortunately for the model, the distance of the bug from the beginning of the program proved a better predictor of debugging times than did the assumed depth of the bug in the hierarchy. However, there was a significant effect of depth of the bug on the probability of detecting it.

Shneiderman and Mayer (1979) propose a cognitive model of programming that is based on the cognitive model of Feigenbaum (1970). According to Shneiderman and Mayer, sensory information is stored in short-term memory, an internal representation of the program is generated in working memory, and long-term memory stores both syntactic (language-specific) information and semantic (problem-solving strategy) information. While the details of the memory structures and processes employed in this model need to be revised (see Anderson & Bower, 1973, chapter 4 for a criticism of Figenbaum's model), the concepts of an internal representation of the program and the distinction between semantic and syntactic knowledge are useful.

Mayer (1976, 1981) demonstrates several techniques for improving the integration of knowledge for students who are learning to program. For example, Mayer (1976) describes the effects of a schematic illustration and the "transactional" level of description of a computer's actions (see also DuBoulay, O'Shea, & Monk, 1981; Mayer, 1981). The use of the illustration during learning improves performance on problems that require creative inferences, however it degrades performance on tasks that involve simple retention. Moreover, the illustration is only beneficial when when it is used during training, and not when it is presented following training. As a general account these findings Mayer relates them to Greeno's (1973) schema model of learning and problem solving. Specifically, Mayer argues that the illustration promotes "rich external connections between new material and existing knowledge."

Brooks (1977) models programming with a production-system simulation. In particular, Brooks describes the way one programmer analyzed and coded segments of four problems in FORTRAN. The simulation consists of 73 productions (rules) which are divided into three groups: rules for overall goal management, rules for outputting code, and rules for generating code. Extrapolating, "tens or hundreds of thousands" of rules would be required to model a programmer's knowledge, but Brooks argues that this is a plausible value because programming is a highly complex activity. Another way of evaluating this model compares aggregated aspects of its performance with analogous aspects of the programmer's performance. Thus, Brooks finds that the number of statements executed by the simulation program is roughly correlated with the number of lines in the programmer's verbal protocol. While it remains uncertain whether the details of the cognitive processes described by the model are accurate, this model seems to be a reasonable first step toward understanding programming in terms of problem-solving activities.

2.2.2 Natural-Language Programming. Natural-language programming is closely related to both human problem solving and natural-language interfaces. As in the case of natural-language interfaces, natural-language programming has both proponents (Biermann & Ballard, 1980; Halpern, 1966; Sammet, 1966) and critics (Dijkstra, 1963, 1964).

In an experimental approach, Miller (1974) asked subjects to specify a procedure for sorting names. Nearly 67% of the errors in this task concerned conditional statements, with "or" relationships and negation being particularly troublesome. Of course, these problems are analogous to the problems with conditionals found for data-base access languages.

Miller (1981) describes a study in which subjects used unconstrained natural language to specify procedures for manipulating files in a simulated information filing and retrieval task. Some of the same types of observations reported for studies of natural-language interfaces are confirmed by Miller, for instance the subjects did not use an extensive vocabulary, only 610 unique words were found in the 84 protocols. In addition, the pervasiveness of contextual referencing (e.g., with pronouns) was noted. The approach of the subjects using natural language may be contrasted to the approach that would be employed in typical programming languages. The natural-language solutions almost never included variable declarations, data type specification, or dimensioning of arrays. Moreover, natural language almost always followed a linear structure with branching only by if-then statements. In addition, the subject frequently omitted details in the sequence of actions to be executed. Thus, transfer of control statements were infrequent and often abbreviated when they did occur. Overall, it seems that full natural-language programming is impractical given the number and complexity of inferences required to interpret the subjects' procedure specifications. However, some natural-language concepts may be incorporated into user-oriented programming languages.

2.3 Problem Solving

Previous sections have discussed various aspects of problem solving; Section 1.2.3 examined production-system theories of problem solving, Section 2.1.1 examined query languages, and Section 2.2 considered cognitive factors in programming; this section is devoted to topics on problem solving not covered earlier.

2.3.1 Computer-Aided Problem Solving. Some researchers argue that a more macroscopic approach, than is typically achieved in information-processing models, is appropriate for understanding the way complex design problems are solved (Thomas & Carroll, 1979; but see Simon, 1969, 1973). Malhotra, Thomas, Carroll, and Miller, (1980) characterize design as a cyclic process of goal elaboration, design generation, and evaluation. By focusing on the processes in each of these stages, tools to aid in problem solving may be specified (Carroll, Thomas, & Malhotra, 1980; Malhotra et al., 1980).

The development of problem-solving systems for managers has been an area of considerable activity but only modest success. For an extensive review see Kanter (1977). The problems may be related in part to the subjective restriction of freedom of action for managers by the systems (Argyris, 1971) and to individual differences in the way information is used (Zmud, 1979).

2.3.2 System Characteristics in Computer-Aided Problem Solving. Some of the earliest studies on the human factors of computer systems involved system characteristics in computer-aided problem solving (Nickerson, 1969). Computer-aided problem solving usually is degraded by both long system-response delays (Goodman & Spence, 1978) and high variability in system-response delays (Miller, 1977). However, for some tasks, delay has the effect of reducing the number of commands issued to the computer without affecting total problem-solution time (Grossberg, Wilson, & Yntema, 1976). In the latter case, apparently, the users spent the delay generating more effective commands.

2.3.3 Computer-Assisted Instruction and Expert Systems. Many aspects of computers make them useful for instruction; for example, the computer may pace teaching to the ability of the student. Indeed, it should be possible to build relatively sophisticated models of the student's knowledge and then to use those models to improve the effectiveness of the training (Carbonell, 1970). Computers may also permit simulation of training situations which might be expensive or even dangerous to provide without the simulation (e.g., Brown, 1977).

Beyond the teaching of students, many "expert systems" are being developed to provide technical advice and support in many professional fields. For a review of several approaches to computer-assisted instruction and expert systems, as well as the educational principles on which they are based, see Hartley (1980).

2.4 Document Preparation

2.4.1 Line-Oriented Versus Screen-Based Editors. Many text editors and formatters have been developed, yet there are only a few studies that compare performance with different editors (Ledgard et al., 1980; Roberts, 1980), or even make specific predictions about their use (Card, Moran, & Newell, 1980b). Thus there has been little analysis of even such fundamental issues as why screen-based editors are usually preferred to line-oriented editors. Among the possible reasons for this preference are that lines of text appear on the screen in the order in which they are to be read, that screen-based editing usually is accomplished by spatial position of a cursor, and that the effects of format manipulations are often directly observable. While there are many apparent advantages to screen-based editors, there may also be disadvantages. For instance, Card et al. (1980b) mention that visually searching for the cursor position on a cluttered screen may be time consuming.

Beyond the features of specific text editors, it is useful to examine the characteristics of the tasks in which editors may be employed. In Section 2.4.2, models for making prespecified edits will be examined; in Section 2.4.3., the use of text editors for impromptu composition and editing will be discussed.

2.4.2 Prespecified Text Edits. Card et al. (1980a) propose the GOMS framework for modeling the use of text editors to make prespecified edits. According to this approach a user steps through a hierarchy of Goals. The goals are accomplished by the action of specific Operators. A set of operators required to accomplish a goal is called a Method. However, since there may be several methods possible to accomplish a goal, the method to be employed in a given task is determined by Selection rules. Card et al. (1980a) examined the utility of this framework by specifying a detailed set of goals, operators, and methods for an editing task for which the POET editor is employed. They then observed that individuals are consistent in the use of selection rules, although there are great differences across individuals. Additional experiments demonstrated that once the selection rules and the duration of operators have been established for an individual, both the sequence of actions and the duration of those actions may be predicted by the model for making prespecified edits.

On one hand, the GOMS approach is impressive since it provides a framework for many different editing processes. Moreover, the models of editing behavior make specific predictions and those predictions are supported with substantial accuracy. On the other hand, the GOMS approach may be criticized because it is only a framework. Thus, it might be argued that the data do not confirm the overall approach as much as they confirm the detailed set of operations proposed for the POET editor.

The Keystroke-Level Model (Card et al., 1980b) describes the time to enter commands on computer systems such as text editors. The time to execute tasks is the sum of the times for the physical actions required to complete the task (such as keystrokes) plus mental time for the user to plan the actions. Like the GOMS Models, the Keystroke-Level Model is restricted to cases where the component activities of the task are known, and to cases in which the user does not make errors and works efficiently. Given these constraints, the Keystroke-Level Model is strongly supported by the data of Card et al. (1980b). However, Roberts (1980) finds that the Keystroke-Level Model's predictions are too small by 25–50% for her data on four different text editors. Clearly, many other factors beyond keystrokes enter into text-editor use, as noted above (Section 2.1.1), the difficulty of recall for command terms can be a major factor in editor use.

2.4.3 Composing and Editing Text. Text-editing programs are also used by authors to prepare and revise their own documents. Obviously, in this mode of editor use, the mental processes associated with text composition are combined with to the processes involved in the use of the editor itself. Some of the theoretical issues in composition processes are discussed in Gregg and Steinberg (1980). For experimental studies of composition, Gould (1978) concluded that letter quality is not greatly affected by composition modality. Although, Gould (1981) found that composing a letter with a text editor is nearly 50% slower than composing letters in handwriting. Although, after adding the time for a secretary to type the handwritten letters, the total time was greater for the handwritten letters.

A detailed study of the times of subsidiary tasks in composition is provided by Allen (1981b). This study reports that there are longer latencies for typing the first character in a word than for other characters. In addition, the mean time for characters in the first word in a phrase is longer than for characters in other words of the phrase. Allen (1982a) has focused on editing of major documents received at a word processing center. The most common type of edit was replacement of one word by another word. Among edits longer than one line, inserting text was very common. Overall, the activities of stylistic correction and elaboration of ideas seemed especially common.

2.4.4 Feedback on Document Content. Starting with programs to check spelling (Peterson, 1980) and more recently with complex aspects of document quality (Cherry, 1982; Macdonald, Frase, Gingrich, & Keenan, 1982), computers are being used to give authors feedback about the content of their documents. One human-factors issue concerns how this often extensive information may be summarized effectively for an author (Macdonald et al., 1982). Another issue is how the availability of this type of information will change the use of text-editing programs, and how the features of text editors may be best integrated with the feedback.

2.4.5 Learning to Use Text Editors. Closely related to the cognitive issues involved in the use of text editors are issues on how people learn to use editing programs. Roberts (1980) systematically compared learning rates for individuals using four different editors. For a set of core-editing operations, Roberts reports a positive correlation across editors between the time to learn the essential operations and the time to perform certain tasks.

Bott (1979) examined how individuals learn text-editor concepts. From protocols of subjects who were attempting to learn editor commands, Bott models the integration of these commands with prior knowledge. The model is based on schemas of organized knowledge, and it suggests that learning occurs by duplicating existing schemas followed by modification of these schemas to accommodate newly acquired information. Thus when modification has not occurred or is incomplete, the default information is that which was associated with the schema in its previous context, and this can lead to errors.

2.4.6 Speech Editing. Just as text editors allow the manipulation of computer-stored character strings, a speech editor manipulates computer-stored speech (Maxemchuk, 1980). In a study of Maxemchuk's speech editor, Allen (1982b) reports that users readily learn to edit their spoken documents; however, the major problem for users is in precisely specifying the points at which edits are to be made.

2.5 Visual Displays

Up to this point, studies dealing with input to a computer have been emphasized; in Sections 2.5 and 2.6, visual and non-visual displays will be examined.

Many aspects of visual displays have been examined in the human-factors literature. Gould (1968) reviews physical considerations for display quality such as luminance, contrast, regeneration rate, and resolution. In addition, there has been considerable recent work (Matula, 1981) on visual fatigue and possible health hazards from the use of Visual Display Units (VDUs). Beyond these considerations, the increased sophistication of displays introduces the need for research on how complex material may be effectively presented and enhanced.

2.5.1 Graphics Displays. VDUs may be used to present line drawings, tables, maps, forms, games, and pictures. In general, research dealing with paper copies of graphs and tables (e.g., MacDonald-Ross, 1977) and with the perception of photographs (Biederman, 1976) should be applicable. Some studies have investigated kinematic (Badre, 1982; Knapp, Moses, & Gellman, 1982) and quasi-spatial (Bolt, 1979) displays, although there is much more to be done in these areas.

The perception of displays must be considered in the context of other mental activities. For instance, Yntema (1963) and Bisseret (1971) emphasize the importance of memory for multiple attributes of displayed objects in air traffic control. Visual displays frequently are employed in interactive contexts such as computer-aided design. Thus many of the issues discussed elsewhere in this paper may be incorporated with graphical displays. For example, the use of command languages for controlling graphical displays has been extensively discussed (Baeker, 1980; Foley & Wallace, 1974). From an even broader perspective, it may be possible to apply some general cognitive principles about the conditions under which visual processing affects task-processing capacity (Kahneman, 1973; Treisman & Davies, 1973).

2.5.2 Text Displays. There are several levels at which information processing involving text displays may be evaluated. For example, visual searches of a text display may be conducted to find a single letter (Gordon & Winwood, 1973; Hirsch, 1981) or word. However, the most common activity involving text displays is reading.

Several aspects of displays affect the ease with which text is read. Legibility of text traditionally has been a concern of applied psychological research (e.g., Bouma, 1980; Tinker 1965). Spatial cues, typography (Coke & Koether, 1981), and color (Engel, 1980) may aid in organization. Of course, a logical structure in the text itself is also important (Wright, 1977).

2.6 Interaction with Computers by Voice and Other Non-visual Modalities

2.6.1 Characteristics of Different Modalities. Voice and writing have traditional areas of application. For example, writing often has been associated with contractual and long-term documentation while voice has been associated with informal interaction (Cochran, Riley, & Stewart, 1980; Limb, 1981). These areas of application are determined in part by the traditional technological constraints of the modalities; for instance, until recently it has been easier to store and manipulate text than voice. Current technology is overcoming many of these traditional limitations and the user's cognitive capabilities in dealing with the information should become more important in the design of systems. Consider the problem of locating a particular statement in a voice letter. If that letter is stored as a tape recording, then locating a statement would require listening to much of the letter at the normal speech rate. However, if that letter is stored on a computer, a wide variety of strategies may be implemented for speeded presentation of the speech and for quickly accessing different sections of the letter. Clearly the effectiveness of different methods depends on the way people can process the information.

Voice and text need not be treated as competing display media; systems combining voice and text could take advantage of the best features of both. Even without capitalizing on the unique capabilities of each sensory modality, the use of two different input channels may be able to reduce attentional demands and improve sensory buffering (Hartman, 1961; Hsia, 1977). For example, voice, text, and even touch can be used as information displays (Burke, Gilson, & Jagacinski, 1980).

The improvement of performance by using different modalities may also be applied to user control of computer systems. For example, typical text editors require text entry and command entry with a keyboard. One common error is that commands are entered as text, and text is entered as commands. It has been argued that this type of error could be minimized by entering commands in voice and text by the keyboard (Limb, 1981).

2.6.2 Speech Recognition and Speech Synthesis. While partial speech recognition can augment existing interface techniques, full speech recognition may transform human-interface design. Both the implementation and implications of a full speech recognition or speech understanding system are closely related to issues in natural-language interfaces. Thus, many of the arguments for and against natural-language interfaces may be applied.

The utility of the limited computer speech recognition currently possible has been studied. Cochran et al. (1980) found that users in a circuit-design task could input approximately 25 isolated words/minute in sustained work-like conditions, while in short bursts of input reached 55 words/minute. Connolly (1979) found for an air traffic control task that voice input of selected words resulted in 64% fewer errors than entering equivalent coded messages on a keyboard, although he found no advantage in rate of entry. While these results seem encouraging, it is necessary to note that limits remain in the tasks and environments to which existing speech recognition systems may be employed.

Studies of the comprehensibility of synthesized speech show little difference from natural human speech (Pisoni & Hunnicut, 1980). Many of the issues in speech synthesis are discussed by Flanagan and Rabiner (1973) and by Michaelis and Wiggins (1982).

3. CONSIDERATIONS FOR THE APPLICATION AND INVESTIGATION OF COGNITIVE FACTORS IN HUMAN INTERFACES

3.1 Cognitive Research and System Design

There should be little doubt that cognitive processes are important for many user interfaces. However, to both enthusiast and skeptic many of the claims must seem bewildering. Several factors contribute to this problem. The complexity of the cognitive processes makes research difficult. Moreover because human cognition is itself an active area of research, many different models and partially supported hypotheses are discussed. Furthermore, for applied work, cognitive processes must be considered in a complex environment with interactions among many processes. Finally, despite the obvious complexity of these issues, nearly everyone seems to have preconceptions about how they think. Thus, there is a great deal of "common sense" about cognitive processes. Intuition can be simply wrong, or perhaps it may be sufficiently accurate to construct the first version of a human interface. At some point, though, a systematic approach should prove more effective than intuition.

3.2 Research Considerations

3.2.1 Research Strategies. There are several quite different strategies that may be employed in systematic research on cognitive ergonomics. The first, which may be called the top-down approach, might attempt to employ one of the several highly developed models for specific applications. In the case of cognitive models predicting complex behavior, this has proven difficult (e.g., Anderson, 1976, p. 535). Of course for more routine tasks, particularly manual tasks, models may be quite accurate (e.g., Card et al., 1980b; Rouse, 1981).

Given the complexity of the issues involved and the varied environments in which systems are used, observation and experimentation are often appropriate. However, different types of experimentation are possible. It is often important to obtain actual usage data for systems so that they may be refined. A related approach is to compare the utility of two specific systems for a particular task. On the other hand, an exploratory experiment may examine general principles of system usage. In any case, an empirical approach to cognitive issues and human interaction with computers must be concerned with the quality of the data about which those inferences are made, and the confidence and generalizability that may be attributed to each conclusion. Methodologists have termed the data quality issue "internal validity" while the generalizability issue has been termed "external validity" (Campbell & Stanley, 1963).

3.2.2 Experimental Methodology: Internal and External Validity. There are many ways in which an experiment may violate internal validity and hence not yield acceptable data. Among the most important considerations are control of environmental factors that may bias the subject, elimination of bias due to the experimenter (Crano & Brewer, 1973), and suitable instrumentation.

There are also many ways that external validity may be violated. One of the most common problems is in generalizations about the subject population. For example, many studies are done with only a few subjects, yet the conclusions are applied to many different users and situations. Even if many subjects are employed, it is necessary to consider whether those subjects reflect the population about which conclusions are to be made (Cuff, 1980). Still another issue for the types of conclusion that can be drawn from an experiment is what the dependent variables are really measuring. This is particularly a problem with respect to cognitive constructs (Bainbridge, 1979; Pachella, 1974).

3.2.3 Experimental Designs, Statistics, and Models. There are accepted standards for designing experiments and analyzing the data obtained. Brooks (1980) has clearly outlined many of the important considerations in experimental design for studies of programming, and many of the same issues are relevant to other areas of research in human interaction with computers. Among the classical experimental designs, Brooks concludes that one of the most effective is one that obtains several different observations from each subject (''within subjects'') while systematically varying experimental conditions (''factorial''). Hand in hand with experimental designs are the statistics applied to confirm conclusions. However, it is well known that statistics sometimes have obscured, rather than illuminated, empirical observations. While statistics are essential to believable inferences, the statistics must be used correctly.

Formal experimental designs and strictly controlled laboratory conditions are well suited to examining many issues in human interaction with computers. However, laboratory studies may not be appropriate for some types of dependent variables (Erdmann & Neal, 1971). Indeed it has been noted that laboratory conditions sometimes may produce artifacts and lead to incorrect conclusions (Chapanis, 1967). One solution is to use quasi-experimental designs (Bair, 1978; Campbell & Stanley, 1963). These are designs that do not meet the standards of formal experiments, but that partially compensate for the lack of experimental control. There are also several different approaches to the analysis of data from nontraditional designs. For example, Tukey (1977) proposes a variety of robust statistics, while Cohen and Cohen (1975) propose a type of regression analysis to allow statistical control of confounded factors.

Many types of models may be used to summarize both data and theories (Johannsen & Rouse, 1979; Rouse, 1981). Of course, the goals of the research affect the type of model employed. Thus regression equations are generally best for quantitative, atheoretical predictions, while more complex models and simulations are useful for many theories.

3.3 Crosscutting Variables

3.3.1 Individual Differences and Experience. It would be of considerable interest if theories could be made about the factors that affect the generalizability of specific experiments. For instance, although one interface may be significantly better than another for most users, there might be a subset of users for whom the other interface would be better (e.g., Gade et al., 1981), and it would be helpful to identify the characteristics of two types of users.

Recently, several studies have examined the effects of experience on the use of computer systems. The approach proposed here is that rather than considering experience as a unitary factor, experience and other individual differences may differentially affect different levels of cognitive processes. For example, in command languages, experience may affect the ability to analyze tasks, the selection of strategies, the reconstruction of command terms, and/or the entry of the commands.

Generalizations about the origin and nature of individual differences, must be made cautiously. Some individual differences may be due to different knowledge and experiences. Other individual differences may be attributed to personality traits. The utility of personality traits has been widely debated among psychologists (Mischel, 1968). Among the major issues are the extent to which traits are stable across time and situations, and the extent to which traits can be accounted for by some global capacity, such as intelligence.

3.3.2 Types of Material. Cognitive processes are often sensitive to the type of stimulus material. For example, recall of information from complex messages is often superior when that material is read rather than heard, while the reverse is true for simple material (Hartman, 1961). In terms of human interaction with computers, different text editors may be optimal for short messages and other text editors would be best for long documents. Likewise, the optimal data-base access method may differ for information that varies in ease of categorization (Allen, 1981a; Durding, et al., 1977) or ease of graphical presentation (Barnard, 1978). While there certainly are differences in materials, there is not yet a simple theory to account for these differences.

3.4 A Complex Milieu

Although research and theories frequently attempt to isolate specific aspects of performance, naturally occurring cognitive processes are embedded in a continuous stream of mental activity. Consider, for example, a person composing electronic mail using a text editor. That person is composing sentences and recalling editor commands, while at the same time, the person may be thinking about a program compilation that was just completed, and be interrupted by a telephone call. Clearly, it is necessary to consider the user's environment as a whole and not simply to optimize various subsystems. Details of performance in such complex situations have rarely been studied and there are only a few general principles. At least in some cases, there are systematic patterns in the way people deal with complex environments (Sperandio, 1971), and with additional study these may be formalized.

Beyond cognitive processes, there are several other important psychological issues in human interaction with computers. For example, computer users' motivation, fears, and social pressures have been widely discussed (e.g., Cooley, 1979; Sheridan, 1980). In the characterization above of a person sending electronic mail consider the possible effects if that same person also were thinking about the impact the mail would have on the recipient, were feeling satisfied from just having eaten lunch, and were anxious a social engagement that evening. Processing may be biased in certain directions; thus, obviously a person will do a job more carefully if there are social or monetary rewards for it. In addition to the direct effects the other psychological factors have for human interaction with computers, they can also combine in complex ways with the cognitive factors. For example, a person in a state of emotional arousal may perform tasks poorly which require complex processing but not change performance on simple tasks (e.g., Rabbitt, 1979).

4. SUMMARY

A great deal of interesting and potentially important work concerning cognitive aspects of human interaction with computers has been reported. Moreover, as computers and software become more sophisticated, the relevance of cognitive factors seems likely to increase. On the other hand establishing satisfactory cognitive models and principles has proven to be an extremely difficult task. The complexity of building suitable models is enormous, perhaps even greater than most researchers believe; however, the benefits from this challenge are probably also very great.

5. ACKNOWLEDGMENTS

This paper benefited from the comments of S. T. Dumais, G. W. Furnas, G. O. Goodman, and T. K. Landauer. This paper will also appear in the journal *Behaviour & Information Technology*.

6. REFERENCES

Abelson, R.P. Psychological status of the script concept. *American Psychologist*, 1981, *36*, 715–729.

Allen, R.B. Cognitive factors in the use of menus and trees: An experiment. *Conference Record NTC' 81*, New Orleans, 1981a, F2.5.1–F2.5.5.

Allen, R.B. Composition and editing of text. *Ergonomics*, 1981b, *24*, 611–622.

Allen, R.B. Patterns of manuscript revisions. *Behaviour & Information Technology*, 1982a, *1*.

Allen, R.B. *Composition and editing of spoken letters*. Bell Laboratories report, 1982b.

Anderson, J.R. *Language, memory, and thought*. Hillsdale, NJ: Lawrence Erlbaum Associates, 1976.

Anderson, J.R., & Bower, G.H. *Human associative memory*. Washington, DC: Winston, 1973.

Argyris, C. Management information systems: The challenge to rationality and emotionality. *Management Science*, 1971, *17*, 275–292.

Atwood, M.E., & Ramsey, H.R. *Cognitive structure in the comprehension and memory of computer programs: An investigation of computer program debugging*. Army Research Institute, 1978.

Badre, A.N. Designing chunks for sequentially displayed information. In A.N. Badre & B. Shneiderman (Eds.), *Directions in human/computer interaction*. Norwood, NJ: Ablex Publishing Co., 1982.

Baeker, R. Towards an effective characterization of graphical interaction. In R.A. Guedj et al. (Eds.), *Methodology of interaction*. Amsterdam: North-Holland Publishing Co., 1980.

Bainbridge, L. Verbal reports of the process operator's knowledge. *International Journal of Man-Machine Studies*, 1979, *11*, 411–436.

Bair, J.H. Productivity assessment of office information systems technology. *Proceedings of the Conference on Trends and Applications: 1978 Distributed Processing*. Gaithersburg, MD, 1978, 12–24.

Ballantine, M. Conversing with computers—the dream and the controversy. *Ergonomics*, 1980, *23*, 935–945.

Barnard, P.J. Task constraints and user variables as determinants of performance in cognitive ergonomics. In J.P. Duncanson (Ed.), *Getting it together: Research and applications in human factors*. Santa Monica, CA: Human Factors Society, 1978.

Barnard, P.J., Hammond, N.V., Morton, J., Long, J.B., & Clark, I.A. Consistency and compatibility in human-computer dialogue. *International Journal of Man-Machine Studies*, 1981, *15*, 87–134.

Biederman, I. On processing information from a glance at a scene: Some implications for a syntax and semantics of visual processing. In S. Treu (Ed.), *User-oriented design of interactive graphics systems*. New York: ACM, 1976.

Biermann, A.W., & Ballard, B.W. Toward natural language computations. *American Journal of Computational Linguistics*, 1980, *6*, 71–86.

Bisseret, A. Analysis of mental processes involved in air traffic control. *Ergonomics*, 1971, *14*, 565–570.

Bolt, R.A. Filing and retrieving in the future—spatial data management? *Infotech state of the art review: Man/computer communication*, Maidenhead, U.K.: Infotech, 1979, 21–36.

Bott, R.A. A study of complex learning: Theory and methodologies (Doctoral Dissertation, University of California, San Diego, 1979). *Dissertation Abstracts International, 1979 39*, 471-B. (University Microfilms No. 7915631)

Bouma, H. Visual reading processes and the quality of text displays. In E. Grandjean & E. Vigliani (Eds.), *Ergonomic aspects of visual display terminals*. London: Taylor and Francis, 1980.

Braine, M.D.S. On the relation between the natural logic of reasoning and standard logic. *Psychological Review*, 1978, *85*, 1–21.

Broadbent, D. E. The magic number seven after fifteen years. In A. Kennedy & A. Wilkes (Eds.), *Studies in long-term memory*. New York: Wiley, 1975.

Broadbent, D.E., & Broadbent, M.H.P. The allocation of descriptor terms by individuals in a simulated retrieval system. *Ergonomics,* 1978, *21,* 343–354.

Brooks, R.E. Towards a theory of cognitive processes in computer programming. *International Journal of Man-Machine Studies,* 1977, *9,* 737–751.

Brooks, R.E Studying programmer behavior experimentally: The problems of proper methodology. *Communications of the ACM,* 1980, *23,* 207–213.

Brosey, M., & Shneiderman, B. Two experimental comparisons of relational and hierarchical database models. *International Journal of Man-Machine Studies,* 1978, *10,* 625-637.

Brown, J.S. Uses of artificial intelligence and advanced computer technology in education. In R.J. Seidel & M.L. Rubin (Eds.), *Computers and communication: Implications for education.* New York: Academic Press, 1977.

Burke, M.W., Gilson, R.D., & Jagacinski, R.J. Multi-modal information processing for visual workload relief. *Ergonomics,* 1980, *23,* 961–975.

Campbell, D.T., & Stanley, J.C. *Experimental and quasi-experimental designs for research.* Chicago, IL: Rand McNally, 1963.

Carbonell, J.R. AI in CAI: An artificial intelligence approach to computer-assisted instruction. *IEEE Transactions on Man-Machine Systems,* 1970, *MMS-11,* 190–202.

Card, S.K., Moran, T.P., & Newell, A. Computer text editing: An information processing analysis of a routine cognitive skill. *Cognitive Psychology,* 1980a, *12,* 32–74.

Card, S.K., Moran, T.P., & Newell, A. The keystroke-level model for user performance time with interactive systems. *Communications of the ACM,* 1980b, *23,* 396–410.

Carroll, J.M., Thomas, J.C., & Malhotra, A. Presentation and representation in design problem solving. *British Journal of Psychology,* 1980, *71,* 143–153.

Chafe, W.L. Some thoughts on schemata. *Proceedings of the Workshop on Theoretical Issues on Natural Language Processing—1.* Cambridge, MA, 1975, 99–101.

Chapanis, A. The relevance of laboratory studies to practical situations. *Ergonomics,* 1967, *5,* 557–577.

Cherry, L. Writing tools. *IEEE Transactions on Communications.* 1982, *COM-30,* 100-105.

Chomsky, N. *Syntactic structures.* The Hague: Mouton and Co., 1964.

Cochran, D.J., Riley, M.W., & Stewart, L.A. An evaluation of the strengths, weaknesses, and uses of voice input devices. *Proceedings of the Human Factors Society Annual Meeting,* Los Angeles, 1980, 190–194.

Cohen, J., & Cohen, P. *Applied multiple regression/correlation analysis for the behavioral sciences.* New York: Wiley, 1975.

Coke, E.V., & Koether, M.E. *Effects of display context on memory for attributive information.* Paper presented at the meeting of the American Educational Research Association, Los Angeles, 1981.

Collins, A.M. The trouble with memory distinctions. *Proceedings of the Workshop on Theoretical Issues in Natural Language Processing—1.* Cambridge, MA, 1975, 56–58.

Connolly, D.W. *Voice data entry in air traffic control.* FAA-NA-79-20, 1979.

Conrad, R. Short-term memory factor in the design of data-entry keyboards: An interface between short-term memory and S-R compatibility. *Journal of Applied Psychology,* 1966, *50,* 353–356.

Cooley, M.J.E. Some social effects of computerism. *Infotech state of the art review: Man/computer interaction,* Maidenhead, U.K.: Infotech, 1979, 55–70.

Crano, W.D., & Brewer, M.B. *Principles of research in social psychology.* New York: McGraw-Hill, 1973.

Cuff, R.N. On casual users. *International Journal of Man-Machine Studies,* 1980, *12,* 163–187.

Davis, R., & King, J. An overview of production systems. In E.W. Elcock & D. Michie (Eds.), *Artificial intelligence 8.* New York: Halsted, 1977.

Dijkstra, E.W. On the design of machine independent programming languages. In R. Goodman (Ed.), *Annual review in automatic programming* (Vol. 3). New York: Pergamon, 1963.

Dijkstra, E.W. Some comments on the aims of MIRAFAC. *Communications of the ACM,* 1964, *7,* 190.

DuBoulay, B., O'Shea, T., & Monk, J. The black box inside the glass box: Presenting computing concepts to novices. *International Journal of Man-Machine Studies,* 1981, *14,* 237–249.

Dumais, S.T., & Landauer, T.K. Psychological investigations of natural command and query terminology. In A.N. Badre & B. Shneiderman (Eds.), *Directions in human/computer interaction.* Norwood, NJ: Ablex Publishing Corp., 1982.

Duncan, J. The demonstration of capacity limitation. *Cognitive Psychology,* 1980, *12,* 75–96.

Durding, B.M., Becker, C.A., & Gould, J.D. Data organization. *Human Factors,* 1977, *19,* 1–14.

Ehrenreich, S. & Porcu, T.A. Abbreviations for automated systems: Teaching operators the rules. In A.N. Badre & B. Shneiderman (Eds.), *Directions in human/computer interaction.* Norwood, NJ: Ablex Publishing Corp., 1982.

Engel, F.L. Information selection from visual display units. In E.J. Grandjean & E. Vigliani (Eds.), *Ergonomic aspects of visual display terminals.* London: Taylor and Francis, 1980.

Erdmann, R.L., & Neal, A.S. Laboratory vs. field experimentation in human factors— an evaluation of an experimental self-service airline ticket vendor. *Human Factors,* 1971, *13,* 521–531.

Feigenbaum, E.A. Information processing and memory. In D.A. Norman (Ed.), *Models of human memory.* New York: Academic Press, 1970.

Feldman, J. Bad-mouthing frames. *Proceedings of the Workshop on Theoretical Issues in Natural Language Processing—1.* Cambridge, MA, 1975, 102–103.

Fitter, M. Towards more "natural" interactive systems. *International Journal of Man-Machine Studies,* 1979, *11,* 339–350.

Flanagan, J.L., & Rabiner.L.R. *Speech synthesis.* Stroudsburg, PA: Dowden, Hutchinson, and Ross, 1973.

Foley, J. D., & Wallace, V. L. The art of natural graphic man-machine conversation. *Proceedings of the IEEE,* 1974, *62,* 462–471.

Gade, P.A., Fields, A.F., Maisano, R.E., Marshall, C.F., & Alderman, I.N. Data entry performance as a function of method and instructional strategy. *Human Factors,* 1981, *23,* 199–210.

Goodman, T., & Spence, R. The effect of system response time on interactive computer-aided problem solving. *Proceedings SIGGRAPH,* Atlanta, 1978, 100–104.

Gordon, I.E., & Winwood, M. Searching through letter arrays. *Ergonomics,* 1973, *16,* 177–188.

Gould, J.D. Visual factors in the design of computer controlled CRT displays. *Human Factors,* 1968, *10,* 359–376.

Gould, J.D. How experts dictate. *Journal of Experimental Psychology: Human Perception and Performance,* 1978, *4,* 648–661.

Gould, J.D. Composing letters with computer-based text editors. *Human Factors,* 1981, *23,* 593–606.

Gould, J.D., & Ascher, R. *Use of an IQF-like query language by non-programmers.* IBM RC 5279, 1975.

Gould, J.D., & Drongowski, P. An exploratory study of computer program debugging. *Human Factors,* 1974, *16,* 258–277.

Green, T.R.G. Programming as a cognitive activity. In H.T. Smith & T.R.G. Green (Eds.), *Human interaction with computers.* London: Academic Press, 1980.

Greeno, J.G. The structure of memory and the process of solving problems. In R. Solso (Ed.), *Contemporary issues in cognitive psychology: The Loyola symposium.* Washington, DC: Winston, 1973.

Gregg, L.W., & Steinberg, E.R. (Eds.), *Cognitive processes in writing.* Hillsdale, NJ: Lawrence Erlbaum Associates, 1980.

Grimes, J.D. A knowledge oriented view of user interfaces. *Proceedings Hawaii International Conference on System Sciences,* Honolulu, 1979, 158–163.

Grossberg, M., Wilson, R.A., & Yntema, D.B. An experiment on problem solving with delayed computer responses. *IEEE Transactions on Systems, Man, and Cybernetics,* 1976, *SMC-6,* 219–222.

Grosz, B.J. Focusing in dialogue. *Proceedings of the Workshop on Theoretical Issues in Natural Language Processing—2.* Urbana, IL, 1978, 96–103.

Halpern, M. Foundations of the case for natural-language programming. *Proceedings Fall Joint Computer Conference,* San Francisco, 1966, 639–649.

Hartley, R. Computer assisted learning. In H.T. Smith & T.R.G. Green (Eds.), *Human interaction with computers.* New York: Academic Press, 1980.

Hartman, F.R. Single and multiple channel communications: A review of research and a proposed model. *Audiovisual Communication Review,* 1961, *9,* 235–262.

Hayes, J.R., & Flower, L.S. Identifying the organization of writing processes. In L.W. Gregg & E.R. Steinberg (Eds.), *Cognitive process in writing.* Hillsdale, NJ: Lawrence Erlbaum Associates, 1980.

Hayes, J.R., & Simon, H.A. Understanding written problem instructions. In L.W. Gregg (Ed.), *Knowledge and cognition.* Potomac, MD: Lawrence Erlbaum Associates, 1974.

Hirsch, R.S. Procedures of the human factors center at San Jose. *IBM Systems Journal,* 1981, *20,* 123–171.

Hsia, H.J. Redundancy: Is it the lost key to better communication? *Audiovisual Communication Review,* 1977, *25,* 63–85.

Johannsen, G., & Rouse, W.B. Mathematical concepts for modeling human behavior in complex man-machine systems. *Human Factors,* 1979, *21,* 733–747.

Just, M.A., & Carpenter, P.A. A theory of reading: From eye fixations to comprehension. *Psychological Review,* 1980, *87,* 329–354.

Kahneman, D. *Attention and effort.* Englewood Cliffs, NJ: Prentice-Hall, 1973.

Kanter, J. *Management-oriented information systems.* Englewood Cliffs, NJ: Prentice-Hall, 1977.

Kelly, M.J., & Chapanis, A. Limited vocabulary natural language dialogue. *International Journal of Man-Machine Studies,* 1977, *9,* 479–501.

Kintsch, W. Comprehension and memory of text. In W.K. Estes (Ed.), *Handbook of learning and cognitive processes* (Vol 6). Hillsdale, NJ: Lawrence Erlbaum Associates, 1978.

Knapp, B.G., Moses, F.L., & Gellman, L.H. Information highlighting on complex displays. In A.N. Badre & B. Shneiderman (Eds.), *Directions in human/computer interaction.* Norwood, NJ: Ablex Publishing Corp., 1982.

Landauer, T.K., Galotti, K.N., & Hartwell, S. *A computer command by any other name: A study of text editing terms.* Bell Laboratories report, 1980.

Ledgard, H.F., Whiteside, J.A., Singer, A., & Seymour, W. The natural larguage of interactive systems. *Communications of the ACM,* 1980, *23,* 556–563.

Lindsay, P., & Norman, D.A. *Human information processing* (2nd ed.). New York: Academic Press, 1977.

Limb, J.O. Integration of media for office services. *Office Automation Conference Digest,* 1981, 353–355.

Macdonald, N.H., Frase, L.T., Gingrich, P.S., & Keenan, S.A. The writer's workbench: Computer aids for text analysis. *IEEE Transactions on Communications,* 1982, *COM-30,* 105–110.

MacDonald-Ross, M. How numbers are shown. *Audiovisual Communication Review,* 1977, *25,* 359–409.

Malhotra, A. *Design criteria for a knowledge-based English language system for management: An experimental analysis.* MIT/LCS/TR-146, 1975.

Malhotra, A., Thomas, J.C., Carroll, J.M., & Miller, L.A. Cognitive processes in design. *International Journal of Man-Machine Studies*, 1980, *12*, 119–140.

Matula, R.E. Effects of Visual Display Units on the eyes: A bibliography (1972–1980), *Human Factors*, 1981, *23*, 581–586.

Maxemchuk, N.F. An experimental speech storage and editing facility. *Bell System Technical Journal*, 1980, *59*, 1383–1395.

Mayer, R.E. Some conditions on meaningful learning for computer programming: Advance organizer and subject control of frame order. *Journal of Educational Psychology*, 1976, *68*, 143–150.

Mayer, R.E. The psychology of how novices learn computer programming. *Computing Surveys*, 1981, *13*, 121–141.

McCracken, D.L., & Robertson, G.G. Editing tools for ZOG, a highly interactive man-machine interface. *Proceedings of the International Conference on Communications*, Boston, 1979, 22.7.1–22.7.5.

Michaelis, P.R., & Wiggins, R.H. A human factors engineer's introduction to speech synthesizers. In A. N. Badre & B. Shneiderman (Eds.), *Directions in human/computer interaction*. Norwood, NJ: Ablex Publishing Corp., 1982.

Miller, G.A. The magical number seven, plus or minus two: Some limits on our capacity for processing information. *Psychological Review*, 1956, *63*, 81–97.

Miller, L.A. Programming by non-programmers. *International Journal of Man-Machine Studies*, 1974, *6*, 237–260.

Miller, L.A. Natural language programming styles, strategies, and contrasts. *IBM Systems Journal*, 1981, *20*, 184–215.

Miller, L.H. A study in man-machine interaction. *Proceedings of the National Computer Conference*, Dallas, June 1977, 409–421.

Minsky, M. A framework for representing knowledge. In P.H. Winston (Ed.), *The psychology of computer vision*. New York: McGraw-Hill, 1975.

Mischel, W. *Personality assessment*. New York: Wiley, 1968.

Monsell, S. Representation, processes, memory mechanisms: The basic components of cognition. *Journal of the American Society of Information Science*, 1981, *32* 378–390.

Moray, N. Where is capacity limited? A survey and a model. *Acta Psychologica*, 1967, *27*, 84–92.

Moray, N. *Workload measurement*. New York: Plenum, 1977.

Newell, A. Harpy, production systems, and human cognition. In R.A. Cole (Ed.), *Perception and production of fluent speech*. Hillsdale, NJ: Lawrence Erlbaum Associates, 1980.

Newell, A., & Simon, H. *Human problem solving*. Englewood Cliffs, NJ: Prentice-Hall, 1972.

Nickerson, R.S. Man-computer interaction: A challenge for human factors research. *Ergonomics*, 1969, *12*, 501–517.

Nickerson, R.S. On conversational interaction with computers. In S. Treu (Ed.), *User-oriented design of interactive graphics systems*. New York: ACM, 1976.

Norman, D.A. Copycat science or does the mind really work by table look up? In R.A. Cole (Ed.), *Perception and production of fluent speech*. Hillsdale, NJ: Lawrence Erlbaum Associates, 1980.

Norman, D.A., & Bobrow, D.G. On data-limited and resource-limited processes. *Cognitive Psychology*, 1975, *7*, 44–64.

Pachella, R.G. The interpretation of reaction time in information-processing research. In B.H. Kantowitz (Ed.), *Human information processing: Tutorials in performance and cognition*. Hillsdale, NJ: Lawrence Erlbaum Associates, 1974.

Peterson, J.L. Computer programs for detecting and correcting spelling errors. *Communications of the ACM*, 1980, *23*, 676–687.

Petrick, S.R. On natural language based computer systems. *IBM Journal of Research and Development*, 1976, *20*, 314–325.

Pisoni, D.B., & Hunnicut, S. Perceptual evaluation of MIT talk: The MIT unrestricted text-to-speech system. *Proceedings of the International Conference on Acoustics, Speech, and Signal Processing,* Denver, 1980, 572–575.

Rabbitt, P.M.A. Current paradigms and models in human information processing. In V. Hamilton & D.M. Warburton (Eds.), *Human stress and cognition: An information processing approach.* Chinchester, U.K.: Wiley, 1979.

Reisner, P. Use of psychological experimentation as an aid to development of a query language. *IEEE Transactions on Software Engineering,* 1977, *SE-3,* 218–229.

Reisner, P. Human factors of data-base query languages: A survey and assessment. *Computing Surveys,* 1981, *13,* 13–31.

Roberts, T.L. Evaluation of computer text editors (Doctoral Dissertation, Stanford University, 1980). *Dissertation Abstracts International 40,* 5338B-5339B. (University Microfilms No. 8011699)

Rouse, W.B. Human-computer interaction in the control of dynamic systems. *Computing Surveys,* 1981, *13,* 71–99.

Rouse, W.B., Rouse, W.B., & Pellegrino, S.J. A rule-based model of human problem solving performance in fault diagnosis tasks. *IEEE Transactions on Systems, Man, and Cybernetics,* 1980, *SMC-10,* 366–376.

Sammet, J.E. The use of English as a programming language. *Communications of the ACM,* 1966, *9,* 228–230.

Schank, R.C., & Abelson, R.P. *Scripts, plans, goals, and understanding.* Hillsdale, NJ: Lawrence Erlbaum Associates, 1977.

Searle, J.R. *Speech acts.* Cambridge: Cambridge University Press, 1969.

Sheil, B.A. The psychological study of programming. *Computing Surveys,* 1981, *13,* 101–120.

Sheridan, T.B. Computer control and human alienation. *Technology Review,* 1980, *83,* 60–73.

Shneiderman, B. *Software psychology: Human factors in computer and information systems.* Cambridge, MA: Winthrop, 1980a.

Shneiderman, B. Natural versus precise concise languages for human operation of computers: Research issues and experimental approaches. *Proceedings of the Association for Computational Linguistics,* Philadelphia, 1980b, 139–141.

Shneiderman, B., & Mayer, R.E. Syntactic/semantic interactions in programmer behavior: A model and experimental results. *International Journal of Computer and Information Sciences,* 1979, *8,* 219–238.

Sime, M.E., Green, T.R.G., & Guest, D.J. Psychological evaluation of two conditional constructions used in computer languages. *International Journal of Man-Machine Studies,* 1973, *5,* 105–113.

Simon, H.A. *The sciences of the artificial.* Cambridge, MA: MIT Press, 1969.

Simon, H.A. The structure of ill structured problems. *Artificial Intelligence,* 1973, *4,* 181–201.

Simon, H.A. How big is a chunk? *Science,* 1974, *183,* 482–488.

Small, D.W., & Weldon, L.J. *The efficiency of retrieving information from computers using natural and structured query languages.* Science Applications Inc., SAI-78-655-WA, 1977.

Sperandio, J.C. Variation of operator's strategies and regulating effects on workload. *Ergonomics,* 1971, *14,* 571–577.

Streeter, L.A., Ackroff, J.M., & Taylor, G.A. *On abbreviating command names.* Bell Laboratories report, 1980.

Thomas, J.C. *Quantifiers and question-asking.* IBM RC 5866, 1976.

Thomas, J.C. A design-interpretation of natural English with applications to man-computer interaction. *International Journal of Man-Machine Studies,* 1978, *10,* 651–668.

Thomas, J.C., & Carroll, J.M. The psychological study of design. *Design Studies,* 1979, *1,* 5–11.

Thomas, J.C., & Carroll, J.M. Human factors in communication. *IBM Systems Journal,* 1981, *20,* 237–263.

Thomas, J.C., & Gould, J.D. A psychological study of Query-by-Example. *Proceedings of the National Computer Conference,* Anaheim, 1975, 439–445.

Tinker, M.A. *Basis for effective reading.* Minneapolis: University of Minnesota Press, 1965.

Tracz, W.J. Computer programming and the human thought process. *Software-Practice and Experience,* 1979, *9,* 127–137.

Treisman, A.M., & Davies, A. Divided attention to ear and eye. In S. Kornblum (Ed.), *Attention and performance IV.* New York: Academic Press, 1973.

Treu, S. Interactive command language design based on required mental work. *International Journal of Man-Machine Studies,* 1975, *7,* 135–149.

Tukey, J.W. *Exploratory data analysis.* Reading, MA: Addison-Wesley, 1977.

Tulving, E. Episodic and semantic memory. In E. Tulving & W. Donaldson (Eds.), *Organization of memory.* New York: Academic Press, 1972.

Watt, W.C. Habitability. *American Documentation,* 1968, *19,* 338–351.

Winograd, T. What does it mean to understand language? *Cognitive Science,* 1980, *4,* 209–241.

Wright, P. Presenting technical information: A survey of research findings. *Instructional Science,* 1977, *6,* 93–134.

Yntema, D.B. Keeping track of several things at once. *Human Factors,* 1963, *5,* 7–17.

Young, R.M. Production systems for modeling human cognition. In D. Michie (Ed.), *Expert systems in a microelectronic age.* Edinburg: University of Edinburg Press, 1979.

Zeigler, B.P., & Sheridan, T.B. Human use of short-term memory in processing information on a console. *IEEE Transactions on Human Factors in Electronics,* 1965, *HFE-6,* 74–83.

Zmud, R.W. Individual differences and MIS success: A review of the empirical literature. *Management Science,* 1979, *25,* 966–979.

2

What Do Novices Know About Programming?*

ELLIOT SOLOWAY
KATE EHRLICH
Department of Computer Science
Yale University
New Haven, Connecticut 06520

JEFFREY BONAR
JUDITH GREENSPAN
Department of Computer and Information Science
University of Massachusetts
Amherst, Massachusetts 01003

We have undertaken a number of studies of novice programmers, in order to better understand the special needs of this growing group. In this paper we review a number of our observations. We first focus on the "plan knowledge" which seems to underly the ability to generate simple looping programs. We then describe the performance of novice and intermediate level programmers on a number of ostensibly simple programming tasks, and try to explain that behavior in terms of the posited underlying knowledge. Based on these analyses, we conclude by emphasizing the importance of teaching more than just the syntax and semantics of a programming language.

1. INTRODUCTION

In a recent book, Shneiderman (1980) points out that:

> "For every professional programmer there are probably ten occasional programmers who write programs for scientific research, engineering development, marketing research, business applications, etc. And finally there are

---

* This work was supported by the Army Research Institute for the Behavioral and Social Sciences, under ARI Grant No. MDA903-80-C-0508 and by the National Science Foundation, under NSF Grant SED-81-12403.

Any opinions, findings, conclusions or recommendations expressed in this report are those of the authors, and do not necessarily reflect the views of the U.S. Government.

> a rapidly growing number of programmer hobbyists working on small business, personal and home computing applications.''

In other words, there will be a great number of people who will program computers in carrying out their daily activities. For such casual programmers, initial difficulties in learning a programming language may become a permanent barrier to their continuing interaction with computers. Clearly, programming languages, text editors, command languages, and other applications programs can be designed to better facilitate human interaction. However, in order to build such systems tailored to the needs of the nonprofessional, we must identify the needs of such individuals. In this volume, there are papers that address themselves to just this issue, e.g., Dumais and Landauer (this volume) explore the factors involved in learnable command languages. Our focus is on programming: what difficulties do nonprofessional programmers have in learning to program, and what are the sources of these difficulties?

In this paper we will present the results of an exploratory study we conducted with nonprofessinal programmers who were asked to write programs as solutions to a set of problems. The objective was to study the *bugs*—errors in programs, and *misconceptions*— misunderstandings in the minds of the novice programmers. We study bugs because they serve to expose the specific knowledge deficiences and/or confusions which people have (Brown & Burton, 1978). The key to understanding the misconceptions of non-professionals is a theory of programming knowledge, i.e., the knowledge which expert programmers have about programming. We assume that misconceptions are some variant of such expert knowledge. As we will soon see, we have developed a preliminary theory of the high-level, plan knowledge which expert programmers appear to have and use. We will describe how this theory guided us in the design of the study reported here, and how it provided the basis for interpreting the bugs in nonprofessionals' programs. In this study, then, we attempt to identify the needs of novice programmers by understanding the source of their difficulties. While not the focus of this paper, we do make suggestions as to how these difficulties might be overcome.

2. AN EXAMPLE

Consider the following problem:

> Write a program which repeatedly reads in integers until their sum is greater than 100. After reaching 100, the program should print out the average of the integers read in.

A program that solves this problem would need a number of components:

1. A way of reading a new integer.
2. A way of accumulating a running total.

3. A way of counting the number of integers read, in order to compute the average.
4. A loop which repeatedly performs the above operations, and in addition has a terminating condition to prevent numbers being read in indefinitely.
5. A way to calculate the average after the appropriate number of integers have been read in.
6. A way to print out the final answer.

A Pascal program which correctly solves this problem is given in Figure 1. Although this simple example illustrates many basic notions of programming, such as looping, testing and operating on variables, and reading/writing, the above problem should not be difficult, even for novices. It requires no esoteric language constructs, nor does it require particularly clever code. However, when we gave this and similar problems to students in an introductory Pascal programming class, only 44% of them were able to write correct programs.

```
program  Example;
     var   Count, Sum, Next : integer;
           Average : real;
     begin
     Count : 0;
     Sum : 0;
     repeat
          Read (Next);
          Sum : Sum + Next;
          Count : Count + 1
     until Sum > 100;
     Average = Sum / Count;
     Writeln ( 'The average is : ', Average)
     end.
```

FIGURE 1. A Sample Program. This program calculates the average of numbers which are repeatedly read until their sum exceeds 100.

To get some idea of where students were going wrong, we broke the program down into its component parts and scored a subject's program with respect to the correctness of each component part. We then generated a profile of the typical program by integrating the code fragments most commonly written by the subjects for each of the components in the program. The resulting—and correct—program was the one shown in Figure 1. Insofar as this method of analysis gives an accurate composite picture of a novice programmer, it suggests that there is no single component of a Pascal program that is troublesome for the majority of students.

The errors in the students' programs stemmed from two sources: students had difficulty with the syntax and semantics of the various programming language constructs, and they had difficulty determining which constructs to use and how to coordinate them into a unified whole. Examples of the former type include: incorrectly forming assignment statements; explicitly—and wrongly—incrementing the index variable in a Pascal *for* loop, etc. Examples of the latter type include: using an inappropriate loop construct in a problem, failing to initialize variables properly, etc.

Some attention has been paid to identifying and cataloging errors of the first sort (see Gannon, 1978). In contrast, this paper deals primarily with errors of the second sort, i.e., those which reflect a confusion or lack of knowledge about the pragmatic and functional factors involved in using various programming language constructs. In order to identify specific pieces of knowledge which are lacking or not well understood, we will compare the knowledge of expert programmers with that of non-expert programmers. Shneiderman (1976) and Adelson (1981) have shown that experts seem to use high-level knowledge to understand programs, while novices tend to focus on the specific statements employed in a program. Building on this work, it is our goal to identify *specific* knowledge differences between experts and non-experts, and examine how different levels of knowledge affect the performance of non-experts. To this end, we have identified a number of experts' high-level plans, and have used them to guide our empirical work with non-experts; the study described in this paper is the first result of that enterprise.

3. USING EXPERT KNOWLEDGE AS A GUIDE

We believe that experts use more than just knowledge of the syntax and semantics of a programming language when they write programs to solve problems. Expert programmers have and use high-level, plan knowledge to direct their programming activities. A plan is a procedure or strategy in which the key elements of the process have been abstracted and represented explicitly.[1] Faced with a new problem, an expert retrieves plans from his/her knowledge base which have proven useful in similar situations, and then weaves them together to fit the demands of the new problem.

We have identified[2] a set of plans that we believe experts use when solving problems of the sort typically encountered in introductory programming courses (see Table 1). While these problems are relatively simple, their solutions illustrate many of the basic notions of programming. The problem given at the beginning of Section 2, uses a few repetitive actions involving simple arithmetic operations.

---

[1] In this discussion, we will gloss over many technical aspects of the theory of planning. For our purposes, the commonsense notion of ''plan'' is sufficient. The interested reader can follow these issues more carefully in texts such as Nilsson (1980).

[2] The authors of this paper have served as the expert programmers in this enterprise. We are currently conducting group tests and gathering video-taped protocols with a number of other expert programmers in order to evaluate our claims.

TABLE 1
Three Simple Looping Problems

Problem 1. Write a program which reads 10 integers and then prints out the *average*. Remember, the average of a series of numbers is the sum of those numbers divided by how many numbers there are in the series.

Problem 2. Write a program which repeatedly reads in integers until their sum is greater than 100. After reading 100, the program should print out the average of the integers entered.

Problem 3. Write a program which repeatedly reads in integers until it reads the integer 99999. After seeing 99999, it should print out the *correct* average. That is, it should not count the final 99999.

The plan commonly used to solve this problem, the Running_Total Loop Plan, describes the various components needed in a program in order to compute a sum.

In order to be precise about the plan knowledge which expert programmers have, we decided to encode this knowledge in a "formal language." That is, we have borrowed a language for representing knowledge from Artificial Intelligence called "frames" (Minsky, 1975).[3] This style of representing knowledge has two properties which we find particularly appropriate to our task:

1. Plans are not represented as atomic entities, but rather are represented as knowledge packets which have rich internal structure. A plan is encoded as a frame, which has slot types and slot fillers.
2. Plans are explicitly linked together by various types of relationships; these relationships are encoded as types of arcs.

Below, we will expand on the relevant structural properties of the frame representation and also, we will describe the specific content of our theory of programming, which is encoded in the frame representation.

As we indicated above, we have identified and represented a fragment of an expert's programming knowledge; the frames used to represent this knowledge are shown in Figure 2. These frame representations of loop plans are composed of "slot types" (e.g., descriptions, variables) and "slot fillers." The former delimit general properties or aspects of loop plans, while the latter are specific values which capture the individual characteristics of a particular loop plan. For example, both the Running_Total Loop Plan and the Find_Satisfactory Loop Plan have the slot-type "description" but the slot filler for each of these plans is different. In effect, a frame represents a template, which is customized to the particular features of the concept being represented.

---

[3] We have implemented a version of this language in LISP on a computer, and have used the implemented language to actually represent the knowledge described below. This machine-readable knowledge base is used by our intelligent tutoring system, MENO-II (Soloway, Woolf, Rubin, & Barth, 1981c).

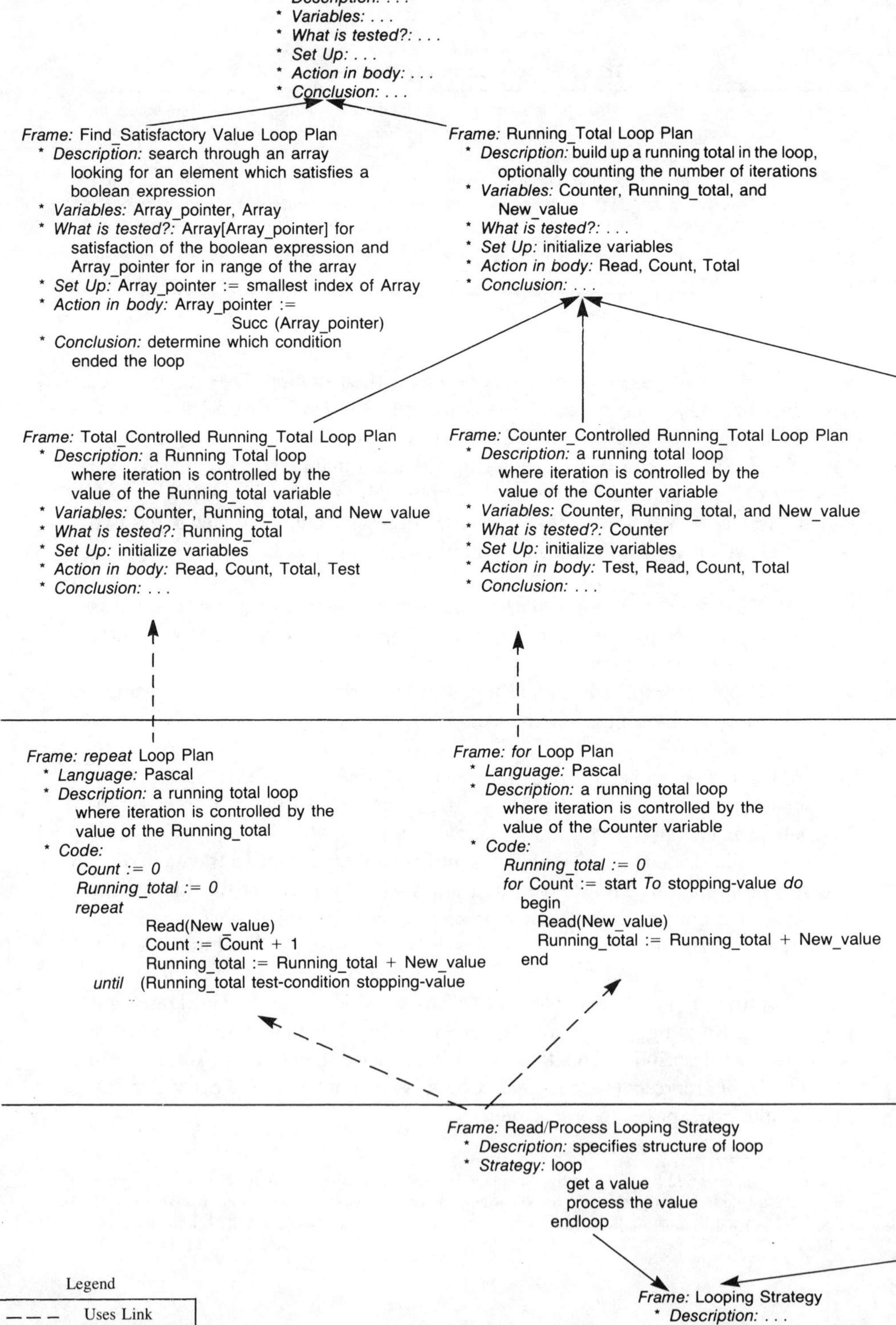

Frame: Loop Plan
* Description: . . .
* Variables: . . .
* What is tested?: . . .
* Set Up: . . .
* Action in body: . . .
* Conclusion: . . .

Frame: Find_Satisfactory Value Loop Plan
* Description: search through an array
    looking for an element which satisfies a
    boolean expression
* Variables: Array_pointer, Array
* What is tested?: Array[Array_pointer] for
    satisfaction of the boolean expression and
    Array_pointer for in range of the array
* Set Up: Array_pointer := smallest index of Array
* Action in body: Array_pointer :=
                Succ (Array_pointer)
* Conclusion: determine which condition
    ended the loop

Frame: Running_Total Loop Plan
* Description: build up a running total in the loop,
    optionally counting the number of iterations
* Variables: Counter, Running_total, and
    New_value
* What is tested?: . . .
* Set Up: initialize variables
* Action in body: Read, Count, Total
* Conclusion: . . .

Frame: Total_Controlled Running_Total Loop Plan
* Description: a Running Total loop
    where iteration is controlled by the
    value of the Running_total variable
* Variables: Counter, Running_total, and New_value
* What is tested?: Running_total
* Set Up: initialize variables
* Action in body: Read, Count, Total, Test
* Conclusion: . . .

Frame: Counter_Controlled Running_Total Loop Plan
* Description: a running total loop
    where iteration is controlled by the
    value of the Counter variable
* Variables: Counter, Running_total, and New_value
* What is tested?: Counter
* Set Up: initialize variables
* Action in body: Test, Read, Count, Total
* Conclusion: . . .

Frame: repeat Loop Plan
* Language: Pascal
* Description: a running total loop
    where iteration is controlled by the
    value of the Running_total
* Code:
    Count := 0
    Running_total := 0
    repeat
            Read(New_value)
            Count := Count + 1
            Running_total := Running_total + New_value
    until   (Running_total test-condition stopping-value

Frame: for Loop Plan
* Language: Pascal
* Description: a running total loop
    where iteration is controlled by the
    value of the Counter variable
* Code:
    Running_total := 0
    for Count := start To stopping-value do
    begin
        Read(New_value)
        Running_total := Running_total + New_value
    end

Frame: Read/Process Looping Strategy
* Description: specifies structure of loop
* Strategy: loop
            get a value
            process the value
        endloop

Frame: Looping Strategy
* Description: . . .
* Strategy: . . .

Legend
- - -   Uses Link
______   A Kind-of-Link

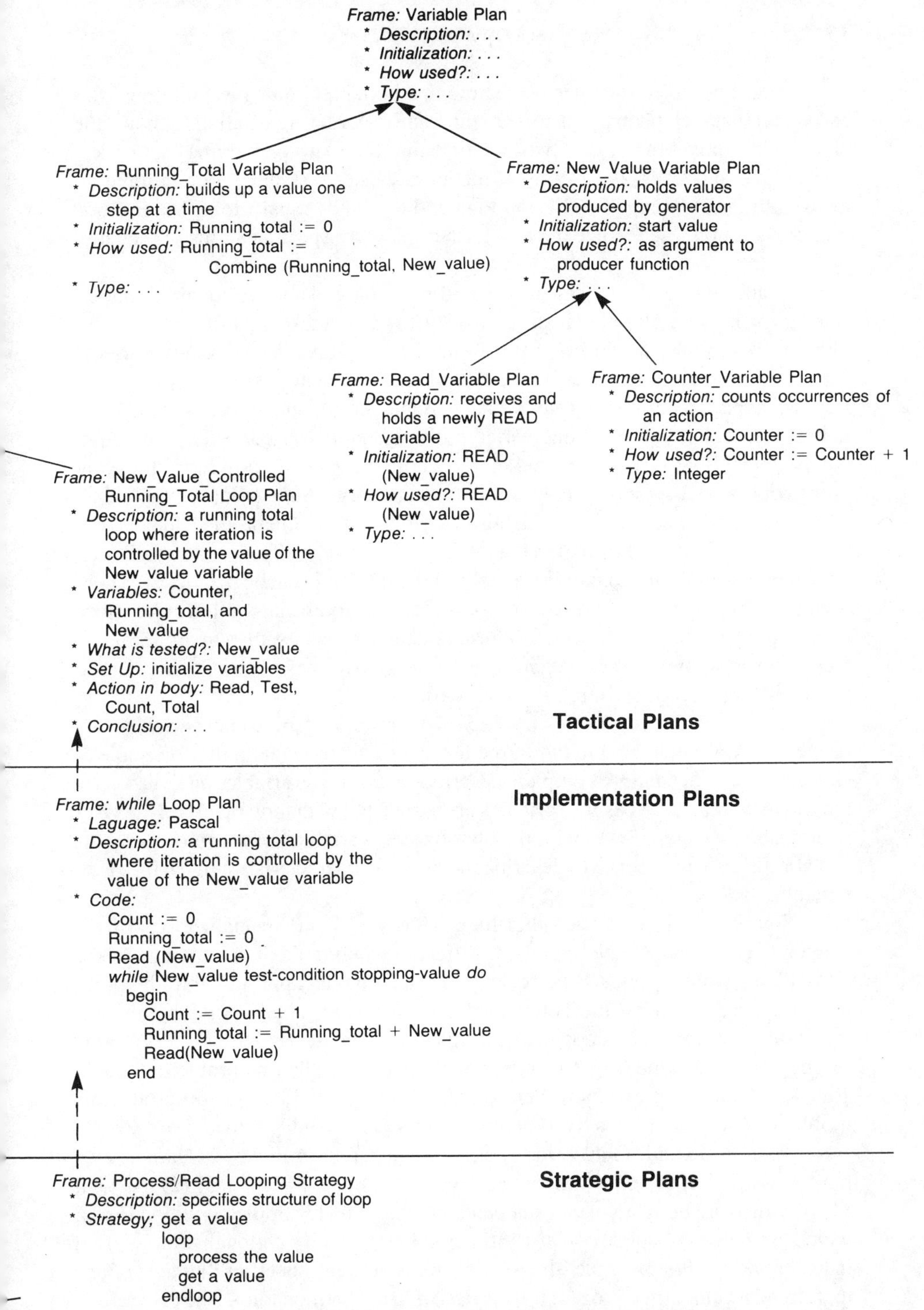

FIGURE 2. A Theory of Simple Looping Programs. The *frames* in this network represent the knowledge which experts seem to have about simple looping problems. Implementation plans here are tailored to Pascal. In order to minimize crossovers, we have not shown Variable Plans being used by Looping Plans.

Plans are linked together—a feature that both highlights the commonalities and differences of plans, and reflects hierachical ordering relations among the plans. Currently, there are two relationship links in our frame network: A_Kind_Of and Uses. For example, the hierarchical A_KIND_OF relation connects both the Running_Total Loop Plan and the Find_Satisfactory_Value Loop Plan to the Loop_Plan, since the two former plans are specializations of the latter plan.

In addition to identifying abstract programming plans, we sought to understand the roles variables play in programs. Just as actors take on different roles in a play, variables take on different functions in a program. We have again used a frame representation to describe the knowledge associated with variable plans. Figure 2 depicts a number of roles which variables commonly play in simple looping problems. For example, one variable, the Counter, is used to count the number of elements being accumulated, e.g., Count:=Count+1. Similarly, the Running_Total Variable is used to accumulate a total, e.g., Sum:=Sum+Nu_Value. While both variables are updated using an assignment statement, they do not seem to be perceived, at least by experts, as "simply" variables. That is, a variable can be perceived as simply containing an arbitrary value, or it can be perceived as containing values which have a specific purpose. The roles that we have described reflect the special purposes which values, and their variables, play in programs. In fact, the data we will present suggests that even novices perceive a Counter Variable as different from a Running_Total Variable.

In writing a program using a particular loop plan, some variables are needed; they are indicated in the frame for that particular plan in the "variables" slot. In order to get a more complete description of such a variable, one must go to the frame which describes it. Variables are associated with loop plans via the Uses relationship in Figure 2. As with the plan frames, variable frames consist of slots and slot fillers which serve to describe the defining characteristics of each type of variable.

Figure 2 includes three other loop plans which are specializations of the Running_Total Loop Plan; one major difference among them is the value of the slot called "what is tested." For example, the loop plan appropriate to the problem given in Section 2 is the Total_Controlled Running_Total Loop Plan; the termination condition on the loop in that problem is when the Runnin_Total Variable reaches a certain value (i.e., 100). In contrast, consider the first problem in Table 1 for which the Counter_Controlled Running_Total Loop Plan is appropriate; this problem requests that exactly 10 numbers be read in and summed.

There is an additional layer of structure apparent in Figure 2. Namely, we have classified plans into three categories: (1) Strategic Plans and (2) Tactical Plans, which are both language-independent plans; and (3) Implementation Plans, which are language-specific plans. Strategic Plans specify a global strategy used in an algorithm. For example, the Read/Process Strategy specifies that the actions that "read a value, then process it" are nested in a repetition loop. The program in

Figure 1 illustrates the use of this strategy; the variable Next is first Read, then it is processed in the statement Sum := Sum+Next. It turns out that the Strategic Plans Read/Process and Process/Read are very important, and we will discuss them further in Section 6.3.

Tactical Plans specify a local strategy for solving a problem. For example, the Counter_Controlled Running_Total Loop Plan describes how a sum can be accumulated. This Loop Plan specifies an algorithm for solving a specific problem. Implementation Plans, on the other hand, specify language-dependent techniques for realizing Tactical and Strategic Plans. Thus, the *for*_Loop Plan is a technique for implementing the Counter_Controlled Running_Total Loop Plan in Pascal.

As we stated in the introduction, our goal is to identify the knowledge differences between expert and non-expert programmers. On the basis of the framework outlined in this section, we have designed and analyzed a broad study of the behavior of non-expert programmers. While the knowledge base in Figure 2 is clearly just a beginning, we are pleased and excited by the leverage it has already given us in understanding programming.

4. A DESCRIPTION OF THE EMPIRICAL STUDY

We wanted to test whether or not non-experts have the kind of plan knowledge described above. In particular, we were interested in seeing if non-experts could distinguish the appropriate context in which to use each of Pascal's looping constructs (the *for, repeat,* and *while* loops), i.e., do non-experts know when it is appropriate to use each of the three Pascal implementation loop plans? The problems displayed in Table 1 were constructed *specifically* to test this question. All three problems require that the average of a set of numbers be computed. However, each problem differs with respect to the condition for terminating the loop. Each of the three terminating conditions is implemented in Pascal using a different loop construct. For example, in Problem 1 (Table 1), the value on which the loop will terminate is known beforehand; in this type of situation, the most appropriate Pascal loop construct is the *for* loop (Wirth, 1971). Pascal was specifically designed to minimize redundant primitive commands.

> In view of its intended usage as a convenient basis to teach programming . . . . emphasis was placed on keeping the number of fundamental concepts reasonably small. (Wirth, 1971)

Thus, the choice of the most appropriate loop construct in a given situation is not merely a matter of individual taste, but rather can be based quite squarely on reason.

In Problem 1, the student is asked to write a program which reads in a specific number of integers, 10 in this case, and then computes their average. As argued above, the Pascal loop construct most appropriate for this problem is the *for* loop. A correct solution to this problem using this construct is given in Figure 3.

```
program  Student12_Problem1;
    var   Count, Sum, Next : integer;
          Average : real;
    begin
    Sum := 0;
    for   Count := 1 to 10 do
          begin
          Read (Next);
          Sum := Sum + Next
          end;
    Average := Sum / 10;
    Writeln ( 'The average is : ', Average)
    end
```

FIGURE 3: The Appropriate Use of a *for* Loop. This program calculates the average of 10 numbers input by a user. It is a solution to Problem 1 of Table 1. This program is a minimally edited student's program.

While in Problem 1 the termination condition is a specific number of integers to be read, in Problem 2 it involves the variable which accumulates the sum, i.e., the program should stop reading when this variable is greater than 100. The Pascal loop construct most appropriate[4] to this problem is the *repeat* loop (see Figure 1). In Problem 3, the loop termination condition is a special value (99999 in this case) which, when read as one of the integers, signals a stop. The Pascal loop construct most appropriate to this problem is the *while* construct (Figure 4).[5]

Even though each Pascal loop construct was designed for a specific looping situation, a correct program can usually be written which does *not* use the most appropriate loop construct. In particular, the *while* loop is the most general construct and can be used to simulate the other two. However, using an inappropriate loop construct can become quite burdensome. In this case, the programmer must add extra code in order to compensate for the poor choice of loop construct. For example, Figure 5 shows two programs which correctly solve Problem 3. Although the *while* loop is the most appropriate construct for this problem, the students chose to use the *repeat* construct; in order to follow through with this decision, additional code was required (compare the programs in Figures 4 and 5). The increased complexity of the code caused by the choice of an inappropriate

---

[4] In Section 6.1 we will present more detailed arguments to support this claim.

[5] A better version of this program would also test that some data had been read in, and thus no divsion by zero was occurring. However, very few students actually included this type of test; the program in Figure 4, generated by a student, is typical of observed programs in this regard.

```
program  Student16_Problem3;
         var Count, Sum, Number : integer; Average : real;

         begin
         Count := 0;
         Sum := 0;
         Read (Number);
         while Number <> 99999 do
                 begin
                 Sum := Sum + Number;
                 Count := Count + 1;
                 Read (Number)
                 end;
         Average := Sum / Count;
         Writeln (Average)
         end.
```

FIGURE 4: The Appropriate Use of *while* Loop. This is a stylistically correct solution to Problem 3 to Table 1.

loop construct would presumably increase the opportunity for error, and the data bear this out.

We administered the three problems described Table 1 as a non-credit quiz to two groups of students. The first group, consisting of 31 students in an introductory Pascal programming class, was tested in the final week of their summer session course. The second group included 52 students enrolled in a second course in programming (a data structures course which used Pascal); they were tested in the tenth week of a 16-week semester. The two groups will be referred to as "novices" and "intermediates" respectively.[6]

In using the terms "novice" and "intermediate," we do not mean to imply that the two groups necessarily differ either uniformly or consistently in their ability to program. As Moher and Schneider (1981) point out, people in the same class may differ considerably in their aptitude for programming and thus care must be exercised in evaluating experimental results. Our goal in using people with varying amounts of programming experience is to focus on the levels that people pass through as they learn to program. That is, we are interested more in the kinds of errors that people make, than in which group of people made the errors. Thus, we feel that this type of exploratory study serves to identify a broad range of interesting behaviors which can then be studied using more carefully controlled experimental techniques.

---

[6] In their introductory courses both groups of students were taught and used all three Pascal loop constructs: *while, repeat, for*.

```
a)      program Student6_Problem3;
            var  Number, Count, Total : integer;
                 Finished : boolean;
            begin
            Count := 0;
            Total := 0;
            repeat
                 Read (Number);
                 Finished := (Number = 99999);
                 if not Finished then
                     begin
                     Count := Count + 1;
                     Total := Total + Number
                     end
            until Finished;
            Writeln ('The average is ', Total/Count)
            end.
b)      program Student16_Problem3;
            var Count, Sum, Num : integer; Average : real;
            begin
            Count := -1;
            Sum := 0;
            repeat
                      Count := Count + 1;
                      Read (Num);
                      Sum := Sum + Num
            until Num = 99999;
            Sum := Sum - 99999;
            Average := Sum / Count
            end.
```

FIGURE 5: Simulating Read/Process with a *repeat* Loop. These two programs are solutions to Problem 3 (Table 1), the *while* loop problem. These programs are minimally edited students' program.

Scoring this type of production data is difficult.[7] Based on the programming knowledge described earlier, we developed a number of categories for coding the students' programs. For the most part, these categories reflected the functional characteristics of the program. For example, we scored a program with respect to

---

[7] Since students wrote the solutions on paper without access to a computer, we felt that it would be counterproductive to score as incorrect, programs which had only minor syntactic errors. Also, we did not mark as incorrect, programs in which there was no check for the count being zero before attempting to divide by it in the average calculation. We made this decision since only a small handful of programs did make the test.

the use/misuse of the Counter Variable, Running_Total Variable, etc. (see Figure 2). This approach goes beyond more straightforward measures such as scoring a program with respect to the use/misuse of a programming language construct (e.g., did the student correctly use an assignment statement?). In effect, the latter technique would simply access the student's understanding of the syntax and semantics of the programming language constructs which, while undeniably important, are not the only components of programming knowledge. For example, if we had just looked at assignment statement usage, we would have missed the distinction between the Counter Variable and the Running_Total Variable. Our goal in developing scoring categories based on programming plans, was to tap into the students' understanding of the functional characteristics of constructs, as well as their understanding of the syntax and semantics of those constructs.

5. OVERALL PERFORMANCE

Table 2 displays the percentage of novices and intermediates who gave a correct solution to the problems. Unquestionably, the figures are low. Novices scored less than 50% on every problem, while the intermediates scored less than 60% on every problem.

TABLE 2
Overall Performance of Novices and Intermediates*

	Problem 1 *for* problem	Problem 2 *repeat* problem	Problem 3 *while* problem
Novices	11/27	12/27	9/23
Correct	(41%)	(44%)	(39%)
Intermediates	29/57	28/49	18/43
Correct	(57%)	(57%)	(42%)

*In a non-credit quiz, novice and intermediate programmers were asked to generate Pascal programs to solve the 3 problems in Table 1. Their performance is summarized in this table.

The majority of errors are not explainable simply as momentary "mental slips," such as failure to initialize a variable. Performance improved only slightly if errors of this sort were overlooked. For example, if we counted as correct all those programs in which the only error was a missing initialization, the percentage of correct programs (across all 3 problems) would only go up by 3% (to 45%) for the novices, and by 6% (to 59%) for the intermediates. Thus, we believe that the errors represented by this data reflect real misconceptions that students have about programming, and cannot be accounted for by simply saying "They made silly errors."

6. BUGS AND MISCONCEPTIONS: LOOPING

As stated earlier, our objective in this study was to build up a catalogue of common programming bugs and their underlying misconceptions. We begin this discussion with a description of bugs and misconceptions concerning the choice of looping construct and looping strategy.

6.1 The Looping Constructs In Pascal

Do novice and intermediate programmers distinguish among the three Pascal loop constructs and use them in the appropriate context? The answer, based on our data, is less than half the time. Table 3 lists the percentage of novices and intermediates who used each particular loop construct on each problem.

Let us first consider the performance of the novice group. Their performance data on Problem 1 are surprising. This problem is clearly a *for* loop problem; the *for* loop was designed for use in situations in which termination of the loop must occur when the Counter Variable exceeds some specific value, e.g., in Problem 1, the Counter Variable must range between 1 and 10. However, only 22% used a *for* loop; 37% used a *repeat* loop and 37% used a *while* loop, even though these constructs required the students to do more work by having to make explicit those operations done implicitly by the *for* loop (i.e., initialize counter, test counter for stopping, increment counter).

Problem 2 (see Table 1) was a *repeat* loop problem; the variable that controlled the loop, "sum," needed to be assigned a value *in* the loop before it could reasonably be tested. Fifty-nine percent of the novices used the *repeat* loop and 37% used the *while* loop; no novices used a *for* loop here. For Problem 3 (Table 1), the appropriate loop construct is *while*; the loop must not be executed if the controlling variable has a specified value and therefore the test must be placed at the head of the loop.[8] Thirty-seven percent of the novices used a *while* loop as opposed to 35% who used a *repeat* loop.

The intermediate group had a different pattern of performance. On Problem 1, the intermediates *did* seem to recognize it as being a *for* loop problem, in contrast to the novice group. However, on Problem 2, in which the *repeat* loop was the most appropriate construct, only a few used it. Rather, the overwhelming choice was the *while* loop. Counter-intuitively, more novices used the appropriate loop construct in this problem than did intermediates. Intermediates again chose the *while* loop in Problem 3. However, given the overall usage of the *while* loop by the intermediates, it appears that the intermediates no longer have the *repeat* loop in their bag of tools; they simply use the *while* loop in all situations. Since the course in which they were tested was a "data structures" course, and did not fo-

---

[8] This, in turn, requires a curious coding structure which we examine in the next section.

TABLE 3
Looping Constructs Used by Novices and Intermediates*

	Problem 1 *for* problem	Problem 2 *repeat* problem	Problem 3 *while* problem
Novices			
Used *for*	6	—	1
	(22%)	(—)	(4%)
Used *repeat*	10	16	8
	(37%)	(59%)	(35%)
Used *while*	10	10	10
	(37%)	(37%)	(37%)
Used other	1	1	4
	(4%)	(4%)	(17%)
Total attempting	27	27	23
Used appropriate			
construct and correct	4/6	7/16	5/10
Used inappropriate			
construct and incorrect	7/21	5/11	4/13
Overall correct	11/27	12/27	9/23
Intermediates			
Used *for*	31	1	—
	(61%)	(2%)	(—)
Used *repeat*	—	5	7
	(—)	(10%)	(16%)
Used *while*	20	38	30
	(39%)	(78%)	(70%)
Used other	—	5	6
	(—)	(10%)	(14%)
Total attempting	51	49	43
Used appropriate			
construct and correct	21/31	3/5	14/30
Used inappropriate			
construct and correct	8/20	25/44	4/13
Overall Correct	29/51	28/49	18/43

*These data present a more detailed view of the data depicted in Table 2.

cus on being tested on Pascal *per se,* they apparently had no motivation to remain familiar with the *repeat* loop; the *while* loop could always be coerced into service.

Based on this simple test, it appears that novices and intermediates do not distinguish among the three Pascal loop structures as experts might. A student we interviewed, when asked why he chose to use the *while* construct rather than one of the other two, responded:

"When I don't know what is going on, I use a *while* loop."

At first, we found the results of the *for* loop problem (Table 2, Problem 1) counter-intuitive. After all, since the *for* loop does so much work automatically, we thought it would be the easiest to understand and use. On second thought, we decided that these automatic, implicit aspects of the *for* loop might in fact be the problem. Since the *for* loop does a number of actions automatically, students might be uncertain about exactly how it works. To gain control over the program, students might choose a *repeat* or *while* loop and do the extra work to add the required looping machinery themselves.[9]

The difference between the *repeat* and *while* loops *is* subtle. Moreover, since one construct can simulate the other, the distinction is hard to enforce. We feel that textbooks significantly contribute to the confusion. For example:

> The principle difference is this: in the WHILE statement, the loop condition is tested *before* each iteration of the loop; in the REPEAT statement, the loop condition is tested *after* each iteration of the loop. (Findlay & Watt, (1978)
>
> If the number of repetitions is known beforehand, i.e., before the repetitions are started, the *for* statement is the appropriate construct to express this situation; otherwise the *while* or *repeat* should be used. . . . The statement [in a *while* body] is repeatedly executed until the expression becomes false. If its value is false at the beginning, the statement is not executed at all. . . . The sequence of statements between the symbols *repeat* and *until* is repeatedly executed (at least once) until the expression becomes true. (Jensen & Wirth, 1974)

As opposed to these ''syntax level'' descriptions, we feel that the ''deep structure'' of these constructions should be emphasized. A better explanation is:

> If the test variable will have a meaningful value as the loop is entered, i.e., a value that could prevent the loop from being executed even once, then a *while* loop is appropriate. If, however, the first meaningful value of the test variable is assigned to it during the loop, then a *repeat* loop is the appropriate iteration construct?

In other words, there is a close connection between the loop test, the loop construct, loop plan, and the problem; the choice of loop construct is dependent on when the test ought to be performed. This observation is consciously reflected in the frame representation of looping plans in Figure 2. The frames and slots in Figure 2 reflect the close connections between the test in the loop and the loop construct, and the connection between the test and the *problem*.

A final comment: the reader might feel that the distinction among the three loop types is not important *for a novice*. To some extent, we might agree. The

---

[9] Some support for this interpretation comes from the following fact: students in these classes were *not* taught the *goto,* and never constructed loops out of more primitive constructs. Thus, they might have been unsure of the ingredients of a loop, and uncomfortable with all the magic implicit in a *for* loop.

problem is that courses and textbooks do teach all three loop types—and students are expected to understand them. Given the above data, however, one wonders if teaching all three is a good strategy.

6.2 Choosing the Appropriate Loop Construct: Predictor of Success? No

Does choosing the appropriate loop construct (i.e., *for, repeat,* or *while* depending on context) for a given problem facilitate the production of a correct program solution? In effect, choosing the appropriate loop construct might "get you into the right ball park." If a student does choose the correct loop construct, then

1. He might know more about programming, and thus be more likely to write a correct program; and/or
2. the loop construct itself might impose constraints that help the student to get the components of the loop correct.

Thus we ask: what was the pattern of performance for those students who *did* choose the appropriate loop construct? Keep in mind that the students in general did not use the loop construct that was appropriate to a problem, and their accuracy was low.

The students (novices and intermediates) who chose the appropriate loop construct were only slightly more likely to write correct programs (see Table 3). In other words, simply choosing the appropriate loop *construct* is not a predictor of program correctness. As we shall describe in the next section, choosing the appropriate looping *strategy* can be a predictor of correctness.

6.3 Choosing the Appropriate Looping Strategy: Predictor of Success? Yes

If the choice of loop *construct* is not a predictor of success then, is the choice of loop *strategy*? Recall from Figure 2 that a looping strategy is a type of Strategic Plan which specifies a global approach to a problem, while a looping construct specifies an Implementation Plan, and is the command in a specific programming language that a programmer selects to realize his strategy. In this section, we will describe two alternative looping strategies, and examine their importance for programming.

We will first focus our attention on the performance pattern of our subjects on Problem 3. While from Table 2 we see that the novices did equally poorly on all 3 problems, the performance of the intermediates is quite different; 57% correctly solved Problems 1 and 2, but only 42% correctly solved Problem 3. While Problem 3 *is* different from Problems 1 and 2, there is no *a priori* reason to expect students to perform any differently than they did on the other problems.

The stylistically correct solution to this problem, using a *while* loop (see Wirth, 1974), requires a curious coding structure:

```
read first-value
while (test ith value)
    process ith value
    read next-ith value
```

The loop must *not* be executed if the test variable has the specified value, and this value could turn up on the first read; thus, a *read* outside the loop is necessary in order to start the loop off. This results in the loop processing being "behind the read;" the ith input is processed and then the next ith $(i + 1)$ is fetched. We call this looping strategy the "process i/read next-i" strategy.

We feel this looping strategy to be unnecessarily awkward and downright confusing (see also Knuth, 1974). A more "natural" looping strategy would be to read the ith value and then process it; we call this latter strategy the "read i/process i" looping strategy. Only through circuitous means can one encode a read i/process i strategy with a *while* loop in problems such as Problem 3. For example, one can embed an *if* test inside the loop which repeats the test in the *while* statement itself, or one can use a Boolean variable and assorted assignments (see Figure 6). Hence, the *while* loop facilitates the coding of a process i/read next-i strategy, but does not facilitate the coding of a read i/process i strategy.

Soloway, Bonar, and Ehrlich (1981a) described an experiment which demonstrated that students do in fact prefer to solve problems using a read i/process i strategy, as opposed to a process i/read next-i strategy. Also, their ability to write correct programs is enhanced if the loop construct facilitates the former strategy. In other words, people do have natural looping strategies, and we need to pay attention to the match between such strategies and the strategies facilitated by programming language constructs.

Given the existence of natural looping strategies, such as read i/process i, and their apparent importance, is the choice of looping strategy a predictor of success? The data in this experiment seem to indicate "yes"—at least for Problem 3. Of the 24 students (novices and intermediates) who used a process i/read next-i strategy, 17 got the program correct, while of the 40 students who used the *while* construct, only 19 got the program correct. For the group choosing the process i/read next-i strategy, the difference been those getting the program correct and those getting it incorrect is significant (Sign test: $N = 24$, $\times = 7$, $p < 0\ .03$), while the corresponding difference for the group who chose the *while* loop is not significant ($N = 40$, $\times = 19$, $p > .10$). Choice of appropriate *strategy* (i.e., process i/read next-i) did predict program accuracy. Choice of appropriate *loop construct* (i.e., *while*) did not predict accuracy.

```
program  Student7_Problem3;
        var N, Sum, X : integer;
            Average : real;
            Stop : boolean;
    begin
    Stop := false;
    N := 0;
    Sum := 0;
    while not Stop do
                begin
                Read (X);
                if X = 99999
                    then Stop := true
                    else begin
                            Sum := Sum + X;
                            N := N + 1
                         end
                end;
    Average := Sum / N;
    Writeln (Average)
    end.
```

FIGURE 6: Simulating Read/Process with a *while* Loop. This program is a correct solution to Problem 3 of Table 1. This program is a minimally edited student's program.

Figure 6 displays a program illustrating the potency of the looping strategies which students bring to a problem—and to a programming language. (Recall our earlier discussion of how a *while* loop can be used to encode a read i/process i strategy with an embedded test.) We also saw more unusual constructions. In Figure 7 we see a program which has a *while* loop embedded in a *repeat* loop.[10] One interpretation is that this student wanted to input a value first, then test it, and, when necessary, jump out of the loop. However, he was not able to marshall the programming constructs to correctly achieve this goal.

We call the kinds of constructions described above (see Figures 6 and 7) "multi-loops." The frequency of multi-loops was not randomly distributed over the three problems. In the novice group, there were 11 "multi-loop" programs, all of which occurred on Problem 3; in the intermediate group, there were also 11 multi-loop programs, all of which occurred on Problem 3. Our interpretation is that students wanted to implement their read i/process next-i strategy, and they went to great lengths to do so in Problem 3.

---

[10] Recall that students in these classes were not taught the *goto* construct; thus we did not see programs which escaped from the middle of the loop via an *if* coupled with a *goto*.

```
program  Student17_Problem3;
    var   Sum, Count, Num : integer;
          Average : real;
    begin
    Sum := 0;
    Count := 0;
    repeat
        Read (Num);
        while Num <> 99999 do
            begin
            Sum := Sum + Num;
            Count := Count + 1
            end
    until Num := 99999;
    Average := Sum / Count;
    Writeln ('The average is ', Average)
    end.
```

FIGURE 7: A "Multi-Loop" Program. This program attempts to solve Problem 3 in Table 1. This student is clearly trying to code the *repeat* loop body as "read/process" and using the embedded *while* as a way to (incorrectly) force an exit from the middle of the *repeat* loop. This program is a minimally edited student's program.

7. BUGS AND MISCONCEPTIONS: VARIABLES

In this section, we will describe a selection of bugs and misconceptions associated with variables found in the novice and intermediate groups' programs. The three types of variables posited earlier (the Read Variable, Counter Variable, and Running_Total Variable) will help to focus this discussion.

7.1 Counter Variable vs. Running_Total Variable: Update Problems

Typically, instructional texts treat the update of the Counter Variable and the update of the Running_Total Variable as instances of the same basic construct, i.e., the assignment statement. If students do treat the two types of variables as the same, we would expect to see no difference in performance, as students should be as likely to handle correctly the Counter Variable update as the Running_Total Variable update. The data tell a different story. On Problem 3, 100% of the novices wrote a correct Counter Variable update, while only 83% wrote a correct Running_Total Variable update. In contrast, the intermediates performed essentially the same on both updates: 91% correct on the Counter Variable update versus 95% correct on the Running_Total Variable update.[11]

Why should the Running_Total Variable update be "harder" than the Counter Variable update for the novices? After all, both require assignment statements of the same form. Below, we list three hypotheses which could account for this observation.

1. The activity of counting and the activity of accumulating a total are *different* activities. Whereas counting is naturally represented as a "successor function" (i.e., increase the previous number by 1), one does not traditionally add up a column of numbers using a running total algorithm. (With the increasing use of pocket calculators, this might change.) However, the standard assignment statement does not distinguish between these two cases, and in fact requires that both activities be encoded in the *same* way. Thus, the mismatch between one's standard algorithm and the programming language results in the Running_Total Variable update being harder to learn than the Counter Variable update.

2. Students might learn the notion of a counter as a special entity, i.e., they see the instructor and the textbook always using $I := I + 1$ whenever a counter is needed, and thus they memorize this pattern as an indivisible unit. In other words, they do not decompose $I := I + 1$ into a left-hand variable which has its value changed by the right-hand expression. By this hypothesis, students are not viewing $I := I + 1$ as an example of an assignment statement.[12] When faced with developing an assignment statement for the "running total function" students must really confront their understanding of the particular type of assignment statement needed in this context. The poorer performance in this situation reflects a misunderstanding of how the assignment statement works, and how to encode a running total with this type of statement.

3. The assignment statement in the Running_Total Variable update case requires a variable while the assignment statement in the Counter Variable update case requires only a constant; the difference in performance reflects the fact that variables are "harder" than constants for novices. While this hypothesis might seem like the obvious one, it unfortunately does not provide an adequate *explanation* for the data. Why should vari-

---

[11] Here we did not include the performance of students on Problems 1 and 2, since Problem 1 focused attention on the Counter Variable and Problem 2 focused attention on the Running_Total Variable.

[12] Performance data on Problem 3 suggest some intriguing corroborating evidence. More novices got the count action correct (e.g., $I := I + 1$) than got the count initialization correct (e.g., $I := 1$)! Maybe this was due to sloppiness. However, if $I := I + 1$ *is* a unit unto itself, then possibly the students do not see the need to initialize the variable.

ables be harder than constants? At the syntactic level, writing down a symbol string does not seem all that much harder than writing down a number. At the semantic level, both are the same—the value of the name is used in the computation. One must go after deeper differences, differences which get at why, and when variables are used as opposed to when constants are used. Hypotheses 1 and 2 above attempt to explain the performance difference by describing the circumstances under which each entity is most appropriate. This hypothesis, while appealing on the surface, lacks the explanatory power that we seek.

Another curious coding technique, which we observed in students' programs, provides additional supportive evidence that students perceive the Running_Total Variable differently from the Counter Variable. The program in Figure 8 contains a Running_Total update which is "fractured," i.e., split over two lines of code (Y:=X+Z, Z:=Y).

```
program  Student21_Problem3;
      var   C, X, Y, Z : integer;
      begin
      C := 0;
      Z := 0
      while X < 99999 do
            begin
            Read (X);
            C := C + 1;
            Y := X + Z;
            Z := Y
            end;
      A := Y div C;
      Writeln (A, ' Average')
      end.
```

FIGURE 8: A Fractured Running_Total Update. This program contains a fractured Running_Total assignment statement in the body of the *while* loop. This program is a minimally edited student's program.

While not incorrect, this technique is not good programming practice. It probably reflects a confusion of the equal sign in traditional algebra with the assignment operator in programming (see Soloway, Lochhead, & Clement 1981b).

If students perceived the Counter Variable update to be of the same sort as the Running_Total Variable update, then we would expect to see "fractured updates" used in both the Counter Variable update and the Running_Total Variable update. Fractured updates were observed in seven programs from the novice group. However, in *all* instances this occurred with respect to a Running_Total

Variable update, not a Counter Variable update. While the numbers are small, and thus conclusions drawn from them must be treated with caution, these data nonetheless are consistent with our claim that students do perceive the two variables as being different.

The observations in this section again illustrate our experimental method: analyze programming knowledge with respect to functional characteristics, rather than simply in terms of the syntactic and semantic characteristics.

7.2 The Read Variable

In all three problems (Table 1), a correct solution required that the program "get a new value with a *read*." Twenty-three percent of all the programs written by novices did not perform this function correctly. Since "read a value" into a variable is a basic technique which is used continually, we were surprised at the high percentage of novices who could not do this at the end of a semester of programming.

Consider, for example, the program depicted in Figure 9, which is typical of programs in the data. This program is an attempt at solving Problem 1. Notice that there is a *read* before the loop on the variable Num, but there is no *read* in the loop. Instead, the variable Num is operated on by the arithmetic operation "subtract 1." What misconception might explain this apparently bizarre program? Recall that in Figure 2 we suggested that the Read Variable and the Counter Variable were both specializations of the New_Value Variable, i.e., both the Read Varia-

```
program  Student19_Problem1;
         var Num, Prev_num, Count, Sum : integer;
         begin
         Count := 0;
         Read (Num);
         Sum := 0;
         repeat
                 Prev_num := Num − 1;
                 Sum := Num + Prev_num;
                 Sum := Sum + 1;
                 Count := Count + 1;
         until Count = 10;
         Average := Sum / Count;
         Writeln
                 ('Average of ten integers is equal to ':2)
         end.
```

FIGURE 9: Overgeneralizing a Counter Variable to a Read Variable. This program is an attempt at Problem 1 in Table 1. This program is a minimally edited student's program.

ble and the Counter Variable were used to hold successively generated values. Possibly, the student who wrote the above program confused a Read Variable with a Counter Variable. Since these variables are "sibling" concepts, this confusion seems quite plausible. Thus, a student might well believe that if subtracting 1 from the value of a Counter Variable will return the previous value of the Counter, then subtracting 1 from the variable Num should also return the previous value of Num. As corroborating evidence, notice that the variable which is assigned the value of Num-1 is called Prev_num.

There is another hypothesis which could account for the difficulty some students seemed to have with the *read*. It could be that students just didn't understand the semantics of the Pascal *read* statement. They may not have perceived that a *read* does two things: it gets a new value, *and* assigns that value to a variable. A language which treated I/O calls as special values that can be assigned "to" a variable or "from" a variable might be more palatable to beginning programmers, e.g.,

```
New_value := Read_from_terminal, or,
Write_to_terminal := Running_sum / Count.
```

Mushed Variables

While experts can distinguish among different kinds of variables, and correctly note when such variables should be used, we found that a substantial percentage of novice programs (27%) used the same variable incorrectly for more than one role. In Figure 10, we display a program with just this sort of "mushed variable" error. In particular, notice that the variable X is used both to store a value being read in {Read (X)} and to hold the running total {X := X+X}. While two distinct variables are needed in order to realize these functions in the program, the program in Figure 10 uses only one variable.

```
program Student26_Problem2;
    var X, Ave : integer;
    begin
    repeat
        Read (X);
        X := X + X
    until X + X > 100;
    Ave := X div Nx;
    Write (Ave)
    end.
```

FIGURE 10: Mushed Variables: An Example. This program is a minimally edited student's program.

Frankly, it is quite difficult to conjure up a simple explanation for why students created such programs. For example, it is possible that a student did recognize that the same variable played two different roles, but assumed that the computer also recognized this and would use the different values appropriately. Often novices impute significant power and wisdom to a computer. We are currently carrying out video-tape interviews with students as they write programs. From this type of data we hope to develop clearer pictures of why students make this type of error.

Only 8% (11/143) of the intermediates' programs contained mushed variables as against 27% (21/77) for the novices. The difference between this value and that for the novices was significant at the $p < .001$ level (chi-square $= 15.44$). One would expect that a student would need to clear up this problem in order to progress in computing. That is, one can understand the intermediates not knowing when to use the *for* and *repeat* loops, but still performing quite successfully with only the *while* loop. Mushing variables, however, is a catastrophic conceptual error which must be overcome in order to proceed.

8. RELATED WORK

Some fine work has been done in Artificial Intelligence and Cognitive Science on understanding problem solving in complex domains. In particular, Collins (1978) has argued for the key role which tacit knowledge plays in the problem solving of experts. Papert's (1980) theory of learning, and his work on the design of LOGO, have impressed upon us the importance of the match between cognitive abilities and programming language constructs. Brown and Burton (1978) and Brown and VanLehn (1980) have developed a sophisticated model to explain the origin of bugs which people exhibit in their subtraction algorithms; the development of a similar model to explain the bugs and misconceptions described here is one of our goals.

Others have mapped out the knowledge which experts use in problem solving in other domains. For example, Polya (1973) and Rissland (1978) have described the knowledge which mathematicians use; they emphasize that mathematician do not actually do mathematics in the "definition, theorem, proof" style employed by most textbooks, but rather use plans and goals to guide their thinking. Larkin, McDermott, Simon, and Simon (1980) have made similar arguments for expert problem-solving behavior in physics. A number of researchers have described various knowledge bases for different areas of programming, e.g., Barstow (1979), Rich (1981), Rich and Shrobe (1978), Miller (1978), and Waters (1979).

Some of the issues we are addressing have also been studied by researchers in the emerging field of "software psychology." For example, Gannon (1978) reports on a set of bugs which he found to occur in student programs. Miller and

Becker (1974) attempted to identify the natural problem solving strategies of non-programmers. Sheppard, Curtis, Milliman, and Love (1979) explore, among other issues, the effect on understanding of different looping structures. Mayer (1980) has explored the utility of providing explicit models as an aid to learning programming. Shneiderman (1976) and Adelson (1981) have done experiments which show that experts tend to see programs in terms of functional units, while novices tend to focus on the individual statements in a program. Shneiderman (1980), in a recent book which brings together much of the research in this field, also emphasizes the important role of higher-level, plan knowledge in programming.

9. CONCLUDING REMARKS

People will need to program with some degree of proficiency in order to continue active participation in our technological society. To facilitate their education, we need to develop programming languages, programming environments, and instructional materials which are tuned to the cognitive abilities of this audience. We are now engaged in building a system which can help students in introductory programming courses to debug their programs and come to grips with their misconceptions (Soloway et al., 1981c). A key component of this project is the empirical work reported here. In order to design and implement material which facilitates learning, we need to understand the sources of the bugs and misconceptions that students have.

In this paper we have examined some of those misconceptions by looking beyond "syntax and semantics" errors to the plans and strategies that people use when they program. By focusing on a few critical aspects of programming, we have shown that looping *strategy* is a key factor in programming. Similarly, we found that students are quite sensitive to the *pragmatic* differences among variables. Finally, while we did not set out in this paper to examine issues pertinent to education, our data leads us to the belief that more emphasis should be placed on teaching the plans and strategies relevant to programming than is done now.

10. ACKNOWLEDGEMENTS

A number of people have read and commented on drafts of this paper: Larry Birnbaum, George Cherry, Ann Drinan, Jim Galambos, Jerry Leichter, G. Michael Schneider, Ben Shneiderman, and Bonnie Webber. We sincerely thank them for their help.

11. REFERENCES

Adelson, B. Problem solving and the development of abstract categories in programming languages. *Memory and Cognition* (Vol. 9.) 1981, 422-433.

Barstow, D. *Knowledge-based program construction*. New York: Elsevier North Holland, Inc., 1979.

Brown, J.S., & Burton, R.R. Diagnostic models for procedural bugs in mathematics. *Cognitive Science*, June 1978, *2* (2), 155–192.

Brown, J.S., & VanLehn, K. Repair theory: A generative theory of bugs in procedural skills. *Cognitive Science*, 1980, *4* (4), 379–415.

Collins, A., *Explicating the tacit knowledge in teaching and learning*. Paper presented at the American Education Research Association, 1978 (also BBN Technical Report 3889).

Findlay, W., & Watt, D. *Pascal: An introduction to methodical programming*. Potomac, MD: Computer Science Press, Inc., 1978.

Gannon, J.D. Characteristic errors in programming languages. *Proceedings of 1978 Annual Conference of the ACM*, Washington, DC, 1978.

Jensen, K., & Wirth, N. *Pascal user manual and report*. New York: Springer-Verlag, 1974.

Knuth, D. Structured programming with *go to* statements. *ACM Computing Surveys*, 1974, *6*, (4), 261–301.

Larkin, J., McDermott, J., Simon, D., & Simon, H. Expert and novice performance in solving physics problems. *Science*, 1980, 208.

Mayer, R. Contributions of cognitive science and related research in learning to the design of computer literacy curricula. In R. Seidel, R. Anderson, B. Hunter (Eds.), *Computer Literacy*, New York: Academic Press, 1982.

Michener E. Understanding understanding mathematics. *Cognitive Science*, 1978 *2*, (4), 361–383.

Miller, M. A Structured planning and debugging environment for elementary programming. *International Journal of Man-Machine Studies*, 1978, *11*, 79–95.

Miller, L., & Becker, C. *Programming in natural english*. (IBM Technical Report RC 5137), November 1974.

Minsky, M. A framework for representing knowledge. In P.H. Winston, (Ed.), *The Psychology of Computer Vision*. New York: McGraw-Hill, 1975.

Moher, T., & Schneider, G. M. A methodology for improving experimentation in software engineering. *IEEE Fifth International Symposium on Software Engineering*, San Diego, CA, 1980.

Nilsson, N. *Principles of artificial intelligence*. Palo Alto, CA, Tioga Publishing Company, 1980.

Papert, S. *Mindstorms, children, computers and powerful ideas*. New York: Basic Books, Inc.

Polya, G. *How to solve it* (2nd ed.), Princeton, N.J.: Princeton University Press, 1973.

Rich, C. A formal representation for plans in the programmer's apprentice. *Proceedings of IJCAI-81*, Vancouver, BC, 1981.

Rich, C., & Shrobe, H. Initial report on a LISP programmer's apprentice. *IEEE Transactions on Software Engineering*, November 1978, *4*, (6),

Sheppard, S.B., Curtis, B., Milliman, P., & Love, T. Modern coding practices and programmer performance. *Computer*, December, 1979.

Shneiderman, B. Exploratory Experiments in Programmer Behavior. *International Journal of Computer and Information Sciences*, 1976, *5* (2), 123–143.

Shneiderman, B. *Software psychology, human factors in computer and information systems*. Cambridge, MA: Winthrop Publishers, Inc., 1980.

Soloway, E., & Woolf, B. Problems, plans, and programs. *Proceedings of Eleventh ACM Technical Symposium on Computer Science Education*, Kansas City, 1980.

Soloway, E., Bonar, J., & Ehrlich, K. *Cognitive factors in programming: an empirical study of looping constructs*. Techn. Rep. 81-10). Amherst, MA: University of Massachusetts, Department of Computer and Information Science, 1981a.

Soloway, E., Lochhead, J., & Clement, J. Does Computer programming enhance problem solving ablility? Some positive evidence on algebra word problems. In R. Seidel, R. Anderson, B. Hunter (Eds.), *Computer Literacy,* New York: Academic Press, 1982.

Soloway, E., Woolf, B., Rubin, E., & Barth, P. Meno-II: An intelligent tutoring system for novice programmers. *Proceedings of IJCAI-81,* Vancouver, BC, 1981c.

Waters, R.C. A Method for Analyzing Loop Programs. *IEEE Trans. on Software Engineering,* May, 1979, SE-5:3.

Wirth, N. The Programming language pascal. *Acta Informatica,* 1971, *1* (1),

Wirth, N. On the composition of well-structured programs. *ACM Computing Surveys,* 1974 *6* (4), 247–259.

3

System Message Design: Guidelines and Experimental Results

BEN SHNEIDERMAN

University of Maryland
Department of Computer Science
College Park, MD 20742

This paper offers informal guidelines for the design of computer system messages. Specificity, constructive guidance, positive tone, user-centered style, and appropriate physical format are recommended as the bases for preparing system messages. These guidelines are especially important when the users are novices, but they can benefit experts as well. Experimental results are presented with COBOL compiler syntax error messages, text-editor messages, and job-control language messages. These results indicate that the phrasing and contents of system messages can significantly impact user performance and satisfaction.

1. INTRODUCTION

User experiences with computer system prompts, explanations, error diagnostics, and warnings play a critical role in influencing acceptance of software systems. Programming languages, command languages, interactive menu selection facilities, and application systems are appreciated not only for the functionality they offer but for the phrasing of system messages in a specific implementation. This paper provides informal guidelines for designing system messages and exploratory experimental results which document the impact of message style on performance and satisfaction. The wording of messages is especially important in systems designed for novice users, but experts benefit from improved messages as well.

Normal prompts, advisory messages, and system responses to commands may influence user perceptions, but the phrasing of error messages or diagnostic warnings is critical. Since errors occur because of lack of knowledge, incorrect understanding or inadvertent slips, the user is likely to be confused, feel inadequate, and be anxious. Error messages with an imperious tone, which condemn the user, can heighten anxiety, making it more difficult to correct the error and increasing the chances of further errors. Messages which are too generic, such as "WHAT?" or "SYNTAX ERROR", or too obscure, such as "FAC RJCT 004004400 400" or "0C7" offer little assistance to the novice user.

These concerns are especially important with respect to the novice user whose lack of knowledge and confidence amplify the stress-related feedback which can lead to a sequence of failures. The discouraging effects of a bad experience in using a computer are not easily overcome by a few good experiences. In fact, I suspect that systems are remembered more for what happens when things go wrong than when things go right. Although these effects are most prominent with novice computer users, experienced users also suffer. Experts in one system or part of a system are still novices for many situations.

Awareness of the difficulties that novices encounter has prompted the development of student-oriented compilers for some languages, which emphasize good diagnostic messages and even limited error correction. The early DITRAN effort (Moulton & Muller, 1967) and CORC (Conway & Maxwell, 1963) were followed by the WATFOR/WATFIV compilers (Cress, Dirksen, & Graham, 1970) and the PL/C compiler (Conway & Wilcox, 1973). These efforts demonstrate what can be accomplished if the developers are sincere about their concern for ease of use. PL/C and WATFIV are widely used in academic environments not only because of their diagnostic messages but also because of their rapid compilation speeds. These systems demonstrate that although there may be a greater development cost for good diagnostics, the production costs can be kept low. Although I am not aware of any controlled experimental research which proves that students using these compilers learn faster, make fewer errors, or have a more positive attitude toward computers, these hypotheses are shared by many people. Rigorous human factors studies would be useful in evaluating the improvement brought about by these systems and would be helpful in convincing skeptics about the importance of designing good system messages.

Producing a set of guidelines for writing system messages is not an easy task because of differences of opinion and the impossibility of being complete (Dwyer, 1981a, 1981b; Golden, 1980). In spite of these dangers, I feel that producing such guidelines could yield better systems. Input parsing strategies, message generation techniques, and message phrasing can be changed without affecting system functionality. More attention to the impact of system messages on users might lead to instrumentation of systems to capture data on error frequency distributions. Such data will enable system designers and maintainers to revise error-handling procedures, improve documentation and training manuals, alter instructional materials, or even change the programming or command language syntax. The complete set of messages should be approved by senior project managers and included in user manuals.

The remaining parts of this section contain a discussion of message attributes with examples of good and bad messages. The second section of this paper presents the results of five exploratory experiments with different message styles.

1.1 Specificity

Messages which are too general make it difficult for the novice to know what has gone wrong. Simple and condemning messages such as "SYNTAX ERROR," "ILLEGAL ENTRY," or "INVALID DATA" are frustrating because they do not provide enough information about what has gone wrong. Improved versions might be "Unmatched left parenthesis," "Legal commands are: Send, Read, File, or Drop," or "Days must be in the range of 1 to 31."

Even in widely appreciated systems like WATFIV, there is room for improvement. Messages such as "INVALID TYPE OF ARGUMENT IN REFERENCE TO A SUBPROGRAM" or "WRONG NUMBER OF ARGUMENTS IN A REFERENCE TO A SUBPROGRAM" might be improved if the name of the subprogram were included and the correct type or number of arguments were provided. The APL system which has so many nice human factors-oriented features comes out poorly when evaluated for system messages. The extremely brief "SIZE ERROR," "RANK ERROR," or "DOMAIN ERROR" comments are too cryptic for novices and fail to provide information about which variables are involved. On the plus side, the standardization (most systems use the APL360 messages) of messages does make it easier for users to move from one system to another. I have long felt that language standardization efforts should include standardization of at least the fundamental messages.

Execution time messages in programming languages should provide the user with specific information about where the problem arose, what variables are involved, and what values were improper. When division by zero occurs, some processors will terminate with a crude message such as "DOMAIN ERROR" in APL or "SIZE ERROR" in some COBOL compilers. PASCAL specifies "division by zero" but may not include the line number or variables that the PLUM compiler offers (Zelkowitz, 1976). Maintaining symbol table and line number information at execution time so that better messages can be generated, is usually well worth the modest resource expenditure.

Systems which offer an error code number leading to a paragraph-long explanation in a manual are also annoying because the manual may not be available and consulting it is disruptive and time consuming. In most cases, system developers can no longer hide behind the claim that printing complete messages consumes too many system resources (Hahn & Athey, 1972; Heaps & Radhakrishnan, 1977).

1.2 Constructive Guidance and Positive Tone

Rather than condemning novice users for what they have done wrong, where possible messages should indicate what they need to do to set things right. Messages such as "DISASTROUS STRING OVERFLOW. JOB ABANDONED." (from a well-known compiler-compiler), "UNDEFINED LABELS," or "ILLEGAL STA. WRN" (both from a major manufacturer's FORTRAN compiler) can be replaced by more constructive phrases such as "String space consumed. Revise program to use shorter strings or expand string space" "Define statement labels before use," or "RETURN I statement cannot be used in a FUNCTION subprogram."

Unnecessarily hostile messages using violent terminology can disturb nontechnical users. An interactive legal citation searching system uses this message: "FATAL ERROR, RUN ABORTED." A popular operating system threatens many users with: "CATASTROPHIC ERROR; LOGGED WITH OPERATOR." There is no excuse for these hostile messages, they can easily be rewritten to provide more information about what happened and what must be done to set things right.

It may be difficult for the compiler writer to write code which accurately determines what the user's intention was, so the advice to be constructive is often difficult to apply. I believe that error-correcting compilers should be extremely conservative for the same reason. Automatic error correction has the danger that users will fail to learn proper syntax, and become dependent on the compiler making corrections for them. Since corrections are compiler dependent, the code and the user's skills many not be portable. For interactive systems, the user can be consulted before corrections are applied.

1.3 User-centered Phrasing

By user-centered I mean that the user controls the system rather than the system directs the user what to do. This is partially accomplished by avoiding the negative and condemning tone in messages and by being courteous to the user. If the system will take a long time to respond to a command then the user should be informed with a simple estimate of the time. Prompting messages should avoid the imperative forms such as "ENTER DATA," and focus on user control such as "READY FOR COMMAND" or simply "READY".

Brevity is a virtue, but the user should be allowed to control the kind of information provided. Possibly the standard system message should be less than a line, but by keying a "?" the user should be able to get a few lines of explanation. Two question marks might yield a set of examples and three question marks might produce explanations of the examples and a complete description. The CONFER teleconferencing system from the University of Michigan provides appealing assistance similar to this. The CDC-PLATO computer assisted instruction system offers a special HELP button to provide explanations when the student needs assistance.

The designers of the Library of Congress' SCORPIO system (Woody, Fitzgerald, Scott, & Power, 1977) for bibliographic retrieval understood the importance of making the users feel that they are in control. In addition to using the properly subservient "READY FOR NEXT COMMAND," the designers avoid the use of the words "error" or "invalid" in the text of system messages. Blame is never assigned to the user but instead the system displays "SCORPIO COULD NOT INTERPRET THE FOURTH PART OF THE COMMAND CONTENTS, WHICH IS SUPPOSED TO BE A 4-CHARACTER OPTION CODE." The message then goes on to define the proper format and present an example of its use.

The phone company, long used to dealing with nontechnical users, offers this tolerant message: "We're sorry, but we were unable to complete your call as dialed. Please hang up, check your number or consult the operator for assistance." They take the blame and offer constructive guidance for what to do. I suspect that a programmer might have generated: "Illegal phone number. Call aborted. Error number 583-2R6.9. Consult your user manual for further information."

1.4 Appropriate Physical Format

Although professional programmers have learned to read upper-case only text, most novices prefer and find it easier to read upper- and lower-case messages. Upper-case only messages should be reserved for serious conditions. Messages that begin with a lengthy and mysterious code number only serve to remind the user that the designers were insensitive to the real needs of users. If code numbers are needed at all, they might be enclosed in parentheses at the end of a message.

There is some disagreement about the placement of messages in a program listing. One school of thought argues that the messages should be placed at the point in the program where the problem has arisen. The second opinion is that the messages clutter the listing and anyway it is easier for the compiler writer to place them all at the end. This is a good subject for experimental study, but I would vote for placing messages in the body of the listing assuming that a blank line is left above and below the message so as to minimize interference with reading the listing. Of course, certain messages must come at the end of the listing, and execution time messages must appear in the output listing.

Some application systems ring a bell or sound a tone when an error has occurred. This can be useful if the error could be missed by the operator, but it is extremely embarrassing if other people are in the room and potentially annoying, even if the operator is alone. The use of audio signals should be under the control of the operator.

The early high-level language, MAD (Michigan Algorithmic Decoder) printed out a full-page picture of Alfred E. Neuman if there were syntactic errors in the program. Novices enjoyed this playful approach, but after they had accumulated a drawer full of pictures, the portrait became an annoying embarrassment. Highlighting errors with rows of asterisks is a common but questionable approach. Designers must walk a narrow path between calling attention to a problem and avoiding embarrassment to the operator. Considering the wide range of experience and temperment in users, maybe the best solution is to offer the user control over the alternatives—this coordinates with the user-centered principle.

1.5 Summary of Recommendations

The wording of system messages may have an impact on performance and attitudes of users, especially novices whose anxiety and lack of knowledge put them at a disadvantage. Improvements might be made by merely using more specific diagnostic messages, offering constructive guidance rather than focusing on failures, employing user-centered phrasing, choosing a suitable physical format, and avoiding vague terminology or numeric codes. These recommendations (Figure 1) emerge from personal introspection and discussions with computer users. Detailed guidelines for constructing improved messages must still be developed and validated through experimental studies with a wide range of users and systems.

```
                        System Messages

        Should Not                |              Should
__________________________________|_________________________________
                                  |
Be                                | Be
 -Wordy                           |  -Brief
 -Negative in Tone                |  -Positive
 -Critical of Errors              |  -Constructive
 -General                         |  -Specific
 -Cryptic                         |  -Comprehensible
Suggest System Control Over       | Emphasize Under Control Over The
             The User             |                          System
                                  |


Other Considerations

        -Upper And Lower Case Is Preferred To Upper Case Only
            Except In Extreme Situations
        -Asterisks Should Be Used Only In Extreme Situations
        -Error Numbers, If Needed At All, Should Be At The End
            Of The Message
        -User Modifiable Message File
        -Two Or More Levels Of Messages
        -Avoid Terms Such as ``ILLEGAL'', ``INVALID'', ``ERROR''
```

FIGURE 1. Summary of system message design guidelines

Many suggestions in the above paragraphs involve merely rewriting the messages, but further gains might be made if designers were willing to modify compilers or other language processors. The inclusion of user-created variable names in compile-time and even run-time messages should contribute to clarity and specificity. This suggestion would require access to the symbol table, but the effort required seems modest. More ambitious modifications might involve changes to the parser to include parses for frequent errors or even more elaborate parsing schemes to help identify the error more accurately. The permissible syntax of command or programming languages might be designed to facilitate error identification; for example, the valid forms might be sparsely distributed over the space of possible inputs. An appealing direction of research would be to create a measure like the Hamming distance of coding theory, which would enable language designers to have a metric of distance between valid syntactic forms.

Other approaches include on-line aids to provide further explanations of messages or examples of correct syntax. For more experienced users, the state of the parse stack, contents of the symbol table, or BNF grammar may be helpful.

2. EXPERIMENTAL RESULTS

The suggestions in Section 1 are merely conjectures. To test the validity of some of these ideas, a series of small exploratory experiments were run Our areas of investigation were messages from a COBOL compiler, text editor, and job-control language processor.

2.1 COBOL Compiler Messages

A pilot study was run to explore the impact of improved messages on the ability of programmers to locate and repair COBOL syntax bugs. The experiment was administered to 22 second-term COBOL students at the University of Maryland in the Fall of 1979.

Five bugs were included in a 132-line COBOL program yielding the following messages from a UNIVAC COBOL compiler:

```
1) RESERVED WORD USED AS PARAGRAPH OR
SECTION NAME IGNORE ATTEMPT RECOVERY HERE
AFTER PREVIOUS ERROR

2) DANGLING ELSE OR WHEN; TREATED AS
AN IMPERATIVE

3) UNDEFINED DATA ITEM STATEMENT OMITTED
ATTEMPT RECOVERY HERE AFTER PREVIOUS ERROR
PREVIOUS ERRORS CAUSE LOSS OF OBJECT CODE
```

```
4) WORD NOT A VERB; SCAN SKIPS TO NEXT VERB
ATTEMPT RECOVERY HERE AFTER PREVIOUS ERROR

5) BLANK MISSING BEFORE OPERATOR OR LEFT
PARENTHESIS BLANK MISSING AFTER ARITH/COND
OPERATOR OR PUNCTUATION
```

A second version of the listing, produced with a text editor, had five improved messages:

```
1) PERIOD IN PREVIOUS LINE CONTAINED IN
IF STATEMENT, DELETE

2) EXTRANEOUS ELSE IN PREVIOUS LINE, DELETE

3) BLANKS IS UNDEFINED DATA ITEM,
MUST USE SPACES

4) USE AFTER PAGE INSTEAD OF AFTER 1 PAGE

5) SPACE REQUIRED BEFORE OPERATOR
SPACE REQUIRED AFTER OPERATOR
```

Code numbers and severity levels were eliminated in the improved messages and a single blank line was left above and below the improved messages. Eleven copies of each of the listings were produced and randomly distributed to the subjects. Seven minutes were allowed to locate and repair the bugs. One point was given for locating the error and two points were given for correcting the bug, yielding a maximum score of 10 points.

The 11 subjects with the UNIVAC COBOL compiler listing had an average of 6.6 points, while the 11 subjects with the improved messages had an average of 8.5 points. A t-test yielded a significant difference at the 5% level.

The results of this pilot study were considered exploratory. It might be difficult to modify a compiler to produce the messages that we used, and in retrospect some of the phrasing seems confusing; for example, it is not clear whether the compiler performed a delete or whether the user must change the statement. Replications should be performed with other messages, professional subjects, and different languages. A more realistic study could be performed if two versions of the same language compiler were available. One group of subjects would be required to work with the standard version and the other group of subjects would work with the improved message version. Capturing performance in actual projects over longer time frames could demonstrate the true impact of improved messages.

2.2 COBOL Compiler Messages: Tone and Specificity

Based on the encouraging results of the pilot study, a more ambitious exeriment was prepared. Our goal in this study was to test the impact of increased specificity and improved tone in COBOL compiler diagnostic messages on debugging. Our hypotheses were that novice programmers would more success-fully correct syntactic errors and give higher subjective evaluations to more specific and non-threatening tone diagnostics.

Subjects. Forty undergraduates taking an introductory COBOL course participated in the experiment. The 14 of these subjects who had taken another programming course earlier were distributed among the four experimental groups. The experiment was administered in the twelth week of a 15-week course to ensure that the subjects had covered all the features that were used in our sample program.

Materials. Nine syntactic bugs were placed in a 116-line COBOL program. The UNIVAC COBOL compiler was used to generate a file containing the program listing with embedded diagnostic messages, and then the text editor was used to make three modified versions of the listing. One listing had improved tone only; for example, instead of the harsh "ILLEGAL REPEAT COUNT ; CHARACTER IGNORED" we used "COMPUTER CANNOT RECOGNIZE REPEAT COUNT CHARACTER; ITEM NOT USED." The third listing had increased specificity only, for example, "NON-NUMERIC PICTURE CLAUSE VALUE; VALUE IGNORED." The fourth listing had both improved tone and increased specificity, for example, "COMPUTER ONLY RECOGNIZES A NUMERIC PICTURE CLAUSE VALUE; CHARACTER NOT USED." While we originally believed that we had generated messages with the kind of improvements that are stated, in retrospect I feel that we could have done better in redesigning the messages according to our criteria. Samples of the messages are in Figure 2.

Copies of the four listings were generated on the line printer so as to resemble normal compiler debugging runs. An experimental consent agreement and a subjective evaluation form were distributed with the listing. The subjective evaluation had only two questions: "How comfortable do you feel with the error messages?" and "How helpful do you feel the error messages were?" Both questions were answered with one of five choices which ranged from "very good" to "very poor."

Administration. The experiment was administered during a normal class session. Participants were told to study the messages on their listing, circle the incorrect syntax, and enter the correct syntax on the listing. All participants were required to remain for the full 15 minutes and answer the subjective evaluation questions at the end.

One point was awarded for correctly circling the error, two points for partial correction, and three points for a complete correction. A perfect score was 27 points.

```
NORMAL TONE/NORMAL SPECIFICITY
------ ------------- ------------
1) VALUE CLAUSE NOT PERMITTED HERE; VALUE CLAUSE IGNORED
2) ROUNDED CLAUSE MEANINGLESS HERE; STATEMENT IGNORED
3) UNDEFINED DATA ITEM; STATEMENT OMITTED
4) OPERAND TOO LARGE; OMIT STATEMENT

NORMAL TONE/IMPROVED SPECIFICITY
------ -------------- -----------
1) VALUE CLAUSE NOT PERMITTED IN FD; VALUE CLAUSE DELETED
2) INCOMING FIELD SMALLER THAN ACCEPTING FIELD; ROUNDED CLAUSE
USELESS; STATEMENT IGNORED
3) UNACCEPTABLE CHARACTER OVER COLUMN 72; ILLEGAL SYNTAX ERROR
4) UNNECESSARY TO HAVE BOTH GIVING/TO OPERAND; ONLY ONE OPERAND
REQUIRED; OMIT STATEMENT

IMPROVED TONE/NORMAL SPECIFICITY
-------- ------------- -----------
1) COMPUTER CANNOT USE A VALUE CLAUSE IN THIS POSITION; CLAUSE
NOT USED
2) ROUNDED CLAUSE MAY NOT APPLY IN THIS INSTANCE, CHECK VALUES;
STATEMENT NOT USED
3) COMPUTER UNABLE TO RECOGNIZE DATA ITEM
4) COMPUTER CANNOT USE EXTRA OPERAND

IMPROVED TONE/IMPROVED SPECIFICITY
-------- -------------- -----------
1) COMPUTER CANNOT USE A VALUE CLAUSE IN A FD; VALUE CLAUSE NOT
USED
2) ROUNDED CLAUSE MAY NOT APPLY WHEN INCOMING FIELD IS SMALLER
THAN ACCEPTING FIELD
3) COMPUTER NOT ABLE TO READ BEYOND COL. 72 CONTINUE TO NEXT LINE
AT COL. 72
4) GIVING CLAUSE NOT NEEDED WHEN TO STATEMENT IS PRESENTED
```

FIGURE 2. Four of the nine messages used in the four versions of the COBOL syntax error messages experiment

Results. The overall average score for the 40 subjects was 13.7, one subject had a perfect score and one subject had a zero score. Table 1 shows the means and standard deviations for the four groups. An analysis of variance showed a statistically significant effect for increased specificity ($p = .074$), but not for improved tone or for the interaction effect. The subjective evaluation question results parallel the performance scores (see Tables 2 and 3) in showing an improvement for increased specificity and a modest improvement for improved tone. The analysis of variance showed a statistically significant difference in the perceived comfort of more specific messages ($p = .074$) and in the helpfulness ($p = .007$). Other effects were not significant.

It was encouraging to note that performance with all three modified listings was superior to performance with the normal compiler listing. This is surprising since the normal messages from the compiler should have been more familiar and well understood by the subjects.

TABLE 1
Means and Standard Deviations for COBOL Syntactic Error Repair Under Different Forms of Compiler Messages. Maximum Score Was 27

	Specificity	
	Normal	Improved
Tone		
Normal	10.7	15.3
	(3.6)	(8.1)
Improved	13.3	15.4
	(3.0)	(6.7)

TABLE 2
Means and Standard Deviations for Subjective Evaluations in COBOL Syntactic Error Repair Study. 1 Indicates Most Comfort with the Compiler Messages, 5 the Least Comfort

	Specificity	
	Normal	Improved
Tone		
Normal	3.4	2.8
	(0.97)	(1.32)
Improved	3.3	2.7
	(0.95)	(0.82)

TABLE 3
Means and Standard Deviations for Subjective Evaluations in COBOL Syntactic Error Repair Study. 1 Indicates Most Helpful with the Compiler Messages, 5 the Least Helpful

	Specificity	
	Normal	Improved
Tone		
Normal	3.2	2.2
	(0.79)	(1.14)
Improved	2.7	2.2
	(0.67)	(0.63)

Discussion. The almost 50% improvement (Table 1) in debugging performance obtained by merely revising the messages demonstrates the importance that wording and content have on novice programmers. The statistically significant difference in performance scores was brought about by improved specificity (Figure 2): using programmer-created variable names, accurately identifying the portion of the statement that was in error, and being more clear about what needs to be done.

Now that we have a clearer idea of how to write diagnostic messages, a further experiment might show more improvement. Although message tone did not have a significant effect in this study, we believe that thought should be given to revising messages which condemn the user and use harsh terminology. User response to specific messages rather than the entire group of nine might clarify this issue. Changing a COBOL compiler should be considered in future studies to create a more realistic environment.

2.3 Presence or Absence of Text-Editor Messages

Text editing is a common task for programmers working on-line and for writers or secretaries preparing various documents. Since the potential user community for text editors is diverse and includes many novice computer users, the design of comprehensible messages is important (Roberts, 1979).

One of the text editors, ED, supplied with the popular UNIX system offers only a question mark if an editor command cannot be interpreted. Since this approach is frustrating to many users (Norman, 1981), simple messages have been added at many installations. The text editors used in this study were ED and a version with one-line messages called BED. Our goal was to see if the addition of simple messages changed performance and satisfaction.

Subjects. The subjects were 21 University of Maryland undergraduate computer science majors who had experience with a line-oriented text editor on the UNIVAC 1100/42. None of the subjects had ever used UNIX.

Materials. A two-page introduction to a subset of the editor commands which were the same for ED and BED, covered INPUT, EDIT, RETYPE, INPUT BEFORE, SUBSTITUTE, MOVE, PRINT, EXIT, and LOCATE.

Two sets of text-editing tasks were constructed with items such as ''In line 2 change REKORD to RECORD'' or ''Locate the line beginning with 'ALTHOUGH' and insert it and the next 3 lines after line 9.'' Set 1 had 12 items and Set 2 had 11 items. A simple final subjective evaluation form was prepared with the same question about both versions: ''On a scale from 0–10, please give your satisfaction with the editor (0 indicates no satisfaction and 10 indicates full satisfaction).''

Administration. Subjects were tested individually or in small groups. Each subject was given eight minutes to study the introductory material. Eleven subjects began working on Set 1 with BED and then repeated Set 1 with ED. Ten subjects began working on Set 2 with ED and then repeated Set 2 with BED. Fifteen minutes was allotted for each set of items and all subjects finished within the time limit. Times were not recorded, only errors were counted. All work was done using a cathode ray tube operating at 300 baud connected to a PDP 11/40. The satisfaction scale was filled in at the end.

Results. The number of errors and the satisfaction scores for the 11 subjects working on Set 1 in the BED-ED order are shown in Table 4. The same dependent measures for the 10 subjects working on Set 2 in the ED-BED order is shown in Table 5. T-tests showed no statistically significant differences in performance or attitude for the BED-ED group. This is not surprising since we expected superior performance with the second editor used. The subjective evaluation scores clearly favored the BED version.

For subjects in the ED-BED group, there was a clear and statistically significant performance superiority with BED ($p < .01$) and subjective preference ($p < .05$). Eight of the 11 subjects in the BED-ED group preferred BED and 9 out of 10 of the subjects in the ED-BED group preferred BED.

Discussion. The results of this repeated-measures design strongly support the hypothesis that for novice users of text editors, simple messages are superior to

TABLE 4
Number of Errors and Subjective Evaluations for Group
That Had BED-ED Sequence With Task Set 1

Subject	Test Scores (Number of Errors)		Subjective Evaluation (0=Low, 10=High Satisfaction)	
	BED	ED	BED	ED
1	8	9	5	3
2	6	6	6	5
3	14	10	1	3
4	11	12	6	3
5	8	10	7	3
6	4	3	5	6
7	21	30	2	0
8	5	9	8	5
9	3	1	6	6
10	10	11	7	8
11	12	8	4	3
Average	9.27	9.91	5.18	4.09
Standard deviation	5.20	7.46	2.14	2.17

TABLE 5
Number of Errors and Subjective Evaluations for Group That
Had ED-BED Sequence With Task Set 2

Subject	Test Scores (Number of Errors)		Subjective Evaluation (0=Low, 10=High Satisfaction)	
	ED	BED	ED	BED
1	23	15	2	7
2	12	4	3	5
3	6	5	5	6
4	9	10	4	4
5	11	8	4	5
6	17	6	1	8
7	4	2	8	9
8	15	4	2	7
9	4	6	8	6
10	6	1	7	8
Average	10.70	6.10	4.40	6.50
Standard deviation	6.22	4.09	2.54	1.58

a single question mark. This may not be surprising, but it must be remembered that the designers of this widely used system did not feel it useful to include simple messages. Similarly, in a popular job-control language processor, user violations of syntax are greeted with an obscure and annoying "WHAT?"

When simple messages are not presented, the user has no way of knowing whether a simple typographic error was made or whether there is a more serious misunderstanding of features. Novices are tempted to simply reenter the same command and become frustrated if they do not succeed. Good messages can reduce anxiety and contribute to learning.

2.4 Tone and Content of Text-Editor Messages

A second experiment on text-editor messages was carried out by Sharon L. Nelson using a paper-and-pencil approach. Our goal in this study was to see if improving the tone and the content of messages would facilitate correction. Pre- and post-test subjective evaluations were used to assess user satisfaction.

Subjects. Subjects for this study were recruited from an introductory FORTRAN course offered by the Department of Computer Science at the University of Maryland. The subjects had three to four months of experience in batch programming mode and none of the subjects had ever used a text editor.

Materials. Three test forms, each seven pages long, were produced. The CONTROL form used messages from the current version of the University of Maryland text editor which runs on the UNIVAC 1100/42. More SPECIFIC-COURTEOUS messages were prepared based on our feeling of what would be a superior message and what could be expected of a redesigned text editor. More HOSTILE-VAGUE messages were prepared to determine if a worse message was more disturbing to the user.

The three test forms had the same layout. Part I was a pretest questionnaire of the subject's attitude towards computers (see Figure 3). Part II was a brief description of the text editor commands TOP, LOCATE, NEXT, CHANGE, and DELETE, with an example of use for each.

Part III, the main part of the experiment, showed a seven-line piece of text with underlined typographical errors. Then ten test items were presented which had an intended correction, the erroneous text-editor command, the system message, and a blank line for the subject to enter the proper command. The 10 messages are shown in all three forms in Figure 4.

Part IV contained a post-test questionnaire of the subject's attitudes toward text-editor messages (Figure 5).

Administration. After a small pilot test, and revisions to instructions, subjects were recruited in groups of 2 to 10. Subjects were given one of the three

```
PRE-TEST QUESTICNNAIRE
-------- --------------

Please answer the following questions:

1. How much experience have you had in computer programming?
   ------------------months

How did you feel during your previous experiences with a
computer?
   2.   FRUSTRATED 7 6 5 4 3 2 1 SATISFIED
   3.   ANGRY 7 6 5 4 3 2 1 PLEASED
   4.   BEWILDERED 7 6 5 4 3 2 1 CONFICENT

Please indicate the choice that best matches your image of a
computer:
   5.   HOSTILE 7 6 5 4 3 2 1 FRIENDLY
   6.   HARMFUL 7 6 5 4 3 2 1 HELPFUL

How would you rate the error messages which you received when you
made mistakes during your previous experiences with computers?
   7.   HOSTILE 7 6 5 4 3 2 1 FRIENDLY
   8.   VAGUE 7 6 5 4 3 2 1 SPECIFIC
   9.   MISLEADING 7 6 5 4 3 2 1 BENEFICIAL
   10. DISCOURAGING 7 6 5 4 3 2 1 ENCOURAGING
```

FIGURE 3. Pre-test questionnaire for text editor messages experiment

```
HOSTILE

 1.    ILLEGAL ENTRY . . . COMMAND REJECTED
 2.    ILLEGAL ENTRY . . . COMMAND REJECTED
 3.    (There is no explicit message here)
 4.    ILLEGAL ENTRY . . . COMMAND REJECTED
 5.    SEARCH FAILURE . . . COMMAND ABORTED
 6.    FATAL ERROR . . . COMMAND REJECTED
 7.    SEARCH FAILURE . . . COMMAND ABORTED
 8.    ILLEGAL ENTRY . . . COMMAND REJECTED
 9.    FATAL ERROR . . . COMMAND REJECTED
10.    INVALID USER REQUEST $ $ $ $ $ $ $ $


CONTROL

 1.    UNDEFINED COMMAND
 2.    SYNTAX ERROR COL. 14
 3.    (There is no explicit message here)
 4.    SYNTAX ERROR COL. 21
 5.    EOF AT LINE 7
 6.    SYNTAX ERROR COL. 44
 7.    STRING NOT FOUND
 8.    UNDEFINED COMMAND
 9.    UNDEFINED COMMAND
10.    NO CURRENT IMAGE


SPECIFIC, COURTEOUS

 1.    CHANE is not a defined command
 2.    LOCATE command must have a blank between command and string
 3.    (There is no explicit message here)
 4.    CHANGE command must have a blank before 1st delimiter
 5.    The remaining text was searched but  CONDITIONC was not found
 6.    Final delimiter is missing from command
 7.    CARDIOVASCULAR is not on the current line
 8.    LOCATE command must have a blank between command and string
 9.    DELETE command must have a numeric argument
10.    Line pointer is pointing to an empty line
```

FIGURE 4. Messages for similar conditions used in text editor messages experiment

forms in random order with an experimental consent agreement. They were told that they had 30 minutes to complete the tasks, that they could ask questions at any time, and that they were to work quietly on their own. No unusual events occurred during administration.

Results. The pretest had nine subjective questions with a 1 to 7 range, so a score of 7 indicated maximum satisfaction and 63 indicated maximum dissatisfaction with computers. Scores for subjects in the HOSTILE-VAGUE, CONTROL, and SPECIFIC-COURTEOUS groups were 35.3, 35.0, and 34.2 respectively. This tight clustering of scores indicates that the initial attitudes to computers and text editors were approximately equal across groups. The task means (Table 6) were 6.4, 4.1, and 7.5 out of a maximum score of 10. The low score for the CONTROL group (4.1) is partially a result of two zero scores from subjects who an-

```
POST-TEST QUESTIONNAIRE
--------- --------------

Please answer the following questions.

How would you rate the error messages which appeared after the
incorrect text editor commands?
     1. HOSTILE 7 6 5 4 3 2 1 FRIENDLY
     2. VAGUE 7 6 5 4 3 2 1 SPECIFIC
     3. MISLEADING 7 6 5 4 3 2 1 BENEFICIAL
     4. DISCOURAGING 7 6 5 4 3 2 1 ENCOURAGING

Would you like to use this text editor?
     5. NEVER 7 6 5 4 3 2 1 OFTEN

How did you feel when trying to understand the text editor
commands and responses?
     6. FRUSTRATED 7 6 5 4 3 2 1 SATISFIED
     7. ANGRY 7 6 5 4 3 2 1 PLEASED
     8. BEWILDERED 7 6 5 4 3 2 1 CONFIDENT
```

FIGURE 5. Post-test subjective evaluation form used in text editor messages experiment

swered the pretest and post-test but did not fill in the task questions. Discarding these tests did not seem appropriate since it might be argued that their scores represented their frustration with these messages. Had we eliminated these scores, the group average would rise to only 5.1. The strong showing, 6.4, by the HOSTILE-VAGUE group might be attributed to the motivating effect of these messages. The best performance was turned in by the SPECIFIC-COURTEOUS group, as expected. T-tests show a significant difference ($p < .02$) only for the CONTROL vs. SPECIFIC-COURTEOUS groups.

TABLE 6
Number Correct (Out of 10) for Three Forms of Text Editor Messages

	Hostile-Vague	Control	Specific-Courteous
	5	3	10
	8	1	10
	8	6	2
	2	6	9
	6	0	8
	1	0	9
	7	5	6
	9	6	9.5
	9	6.5	2
	9	7.5	6.5
Average	6.4	4.1	7.5
Standard deviation	2.9	2.0	2.6

TABLE 7
Subjective Ratings of Messages and Text Editors From Post-Test Questionnaire.
Values Are the Sums of Choices on 1 (Positive Choice Indicating Satisfaction)
to 7 (Negative Term Indicating Dissatisfaction) Scale

	Hostile-Vague	Control	Specific-Courteous
	30	46	14
	25	48	28
	13	30	44
	30	17	26
	39	37	29
	52	38	28
	36	33	35
	29	17	13
	37	18	36
	38	18	20
Average	32.9	30.2	27.3
Standard deviation	10.2	12.1	9.7

The post-test subjective evaluation scores (Table 7) showed mean values in
the expected direction: most positive reaction to the SPECIFIC-COURTEOUS
messages (27.3) and most negative reaction to the HOSTILE-VAGUE messages
(32.9). The CONTROL group had a mean score of 30.2. None of the differences
were statistically significant.

Discussion. The results of this experiment suggest that the wording and
content of text-editor messages make a difference in user performance and attitude. Hostile and vague messages may motivate performance but lead to lower
user satisfaction. More specific and courteous messages can lead to higher performance and satisfaction. We were pleased that a significant difference emerged
in the performance measure, although the small number of subjects and the two
anomalous test papers suggest that a replication is in order.

It is possible that the differences among the messages was more subtle than
we perceived—a repeated measures design in which each subject sees two or three
of the forms might expand the range of subjective evaluations. A replication of
this exploratory experiment with other subjects is being conducted.

The subjective evaluation post-test was designed on an ad hoc basis. Such
scales need to be refined and validated so that they can be used in a variety of experimental and design studies. A reliable scale would be valuable in comparing
versions of a system or competing systems.

2.5 Job-Conrol Language Messages

Another source of annoying system messages is the job-control language
processor. Since job-control languages were designed for expert users, little attention was given to ensuring ease of use. Complex syntactic rules, arbitrary abbrevi-

ations, unforgiving limitations on the use of blanks, rigid positioning rules, and multiple default conditions contribute to the high rate of user errors. Although it is understandable that novices have difficulty with job-control language commands, Mosteller (1981) demonstrates that expert users also have frequent problems. We conducted an experiment to test the effect of variations in the wording of system messages on user comprehension and capacity to repair job-control commands.

Subjects. Sixty-six undergraduates taking an introductory COBOL course participated in this experiment. The students were familiar with UNIVAC-job control language and may have seen some of the messages used in this study. The experiment was administered in the eleventh week of a 15 week course.

Materials. Six comparable system messages were identified on the UNIVAC and IBM MVS systems, and informally designed improvements were generated (Figure 6). For example, the UNIVAC message "OPERATOR KILLED RUN VIA X-KEY IN" and the IBM message "IEE301I CANCEL COMMAND ACCEPTED, IEF45OI YOUR JOB . . . ABEND S222" were contrasted with our improved "THE COMPUTER SYSTEM CONSOLE OPERATOR TERMINATED YOUR JOB." For each message, subjects had to choose the best of four choices, for example:

```
THIS MESSAGE WAS NOT CAUSED BY ERRCRS IN
YOUR CONTROL CARDS. TO DETERMINE WHAT
HAPPENEC, YCU SHOULD:

A) VERIFY THAT THE SYSTEM READ THE LAST
DATA CARD IN YOUR RUNSTREAM.

B) CHECK ALL PARTS OF YCUR PROGRAM THAT USE
THE 'X-222' FACILITY FCR ERRCRS.

C) ASK THE OPERATOR WHY THE JOB WAS
TERMINATED.

D) CHECK THE SOURCE PROGRAM FOR LOGIC ERRCRS.
```

A test item consisted of the system message followed by the task and the four choices. At the end of each item the following subjective evaluation question appeared:

```
Your ease of understanding of this message:
1--2--3--4--5--6--7--8--9--10
EASY                        HARD
```

```
UNIVAC

1.  ADD ELEMENT NOT FOUND
2.  OPERATOR KILLED RUN VIA X-KEY IN.
3.  FAC REJECTED 400010000000
4.  PROGRAM NOT FOUND
    RUNSTREAM ANALYSIS TERMINATED
5.  FAC REJECTED 440440000000
6.  MISSING ACCOUNT


IMPROVED

1.  THE REQUESTED SET OF CARD-IMAGES TO BE INCLUDED IN YOUR RUN
    WERE NOT FOUND UNDER THE NAME YOU SUPPLIED
2.  THE COMPUTER SYSTEM CONSOLE OPERATOR TERMINATED YOUR JOB
3.  THE GROUP OF DATA THAT YOUR RUN REQUESTED WAS NOT FOUND
    (UNDER THE NAME YOU SUPPLIED) TO BE ON THE SYSTEM'S DISK
    STORAGE
4.  AT LEAST ONE OF YOUR CONTROL CARDS COULD NOT BE READ BY THE
    SYSTEM PROGRAM, PROBABLY DUE TO SYNTAX/TYPING ERROR(S)
5.  YOU DID NOT MATCH THE READ PASSWORD ON FILE IN THE COMPUTER
    FOR THE DATA YOU REQUESTED ACCESS TO (YOU PROBABLY SPECIFIED
    INCORRECT PASSWORD)
6.  THE SYSTEM CONTROL PROGRAM COULD NOT LOCATE AN ACCOUNT NUMBER
    ON THE FIRST CARD IMAGE OF YOUR COMPUTER RUN


IBM

1.  IEF612I PROCEDURE NOT  FOUND
2.  IEF301I CANCEL COMMAND ACCEPTED
    IEF450I YOUR JOB . . ABEND S222
3.  IEF212I YOUR.FILE -DATA SET NOT FOUND
4.  IEF452I YOUR JOB NOT RUN - JCL ERROR
5.  IEC117I DATA SET CANNOT BE USED - PASSWORD INVALID
6.  IEF634I ACCOUNT NUMBER MISSING ON THE JOB CARD
```

FIGURE 6. Messages for similar conditiont used in job control message experiment

Booklets containing six test items were prepared for the three versions of the job-control system messages: IBM, UNIVAC, and improved.

Administration. A pilot test with nine subjects was run to refine the test items and to gauge the amount of time necessary for the experiment. A total of 66 students in three sections of the same introductory COBOL course participated during their normal class session. An experimental consent form was signed by each subject and the booklets were handed out so as to alternate among the three versions. Subjects were told that they had eight minutes to answer the questions and the subjective evaluations and that they must stay seated for the full eight minutes. No problems were encountered and the subjects were thanked for their participation after the test booklets were collected.

Results. Grading of the multiple choice items was straightforward— one point for a correct answer, zero for an incorrect answer. The scores and means for each version are given in Table 8.

TABLE 8
Numbers of Subjects Making Correct Choice
(Out of 22) in Comprehension Test of Job Control
Language Messages

Question	UNIVAC	IBM	Improved
1	14	16	21
2	9	8	13
3	8	17	19
4	16	15	20
5	12	15	17
6	5	12	16
Average	10.7	13.8	17.7

Two-tail t-tests revealed a significant difference in performance between the UNIVAC and the improved messages ($p < .001$) and between the IBM and the improved messages ($p = .014$). In both cases the means (out of maximum of 6) favored the improved messages: improved messages (4.8), IBM (3.8), and UNIVAC (2.9).

The subjective evaluations scores are given in Table 9. One indicated the best ease of understanding, and 10 indicated the worst. The improved message mean scores indicated superior ease of understanding: improved messages (3.9), IBM (5.8), and UNIVAC (6.5).

TABLE 9
Average Number of Circled on Ease of
Understanding Question (1=EASY, 10=HARD)

Question	UNIVAC	IBM	Improved
1	4.04	5.45	3.09
2	8.23	9.23	5.72
3	7.86	3.18	4.13
4	5.32	6.00	2.63
5	7.82	7.41	4.54
6	5.45	3.41	3.32
Average	6.45	5.78	3.91

Discussion. Mastering job-control language requires a deep understanding of many hardware and software features and the capacity to recall complex syntactic details. Expert programmers eventually come to master job control but may have difficulty with unfamiliar features or minor errors. Novice programmers, with the need to learn so much detail, are likely to make job-control language errors. This experiment suggests that improving the messages from the job-control language processor may make it easier and more pleasant for novice programmers to get their work done. The obscure terminology, meaningless codes, and occasionally hostile tone appear to interfere with the novice programmers efforts. By using more familiar terms and more descriptive phrasing, it was possible to show higher performance and subjective evaluations.

This experiment needs to be replicated in an on-line environment with actual user errors. In spite of the artificial nature of our experiment, it seems reasonable to suggest that there is room for improvement in the wording and phrasing of job-control language messages. Although our subjects were most familiar with the UNIVAC system and style of messages, their performance with and evaluation of the UNIVAC messages was the worst of the three. It is not surprising that our subjects did poorly on the octal code number message used for UNIVAC item 3, but they did quite well with the octal code number message used in UNIVAC item 5. UNIVAC item 6, ''MISSING ACCOUNT,'' turned out to be the most difficult, possibly because subjects thought the account was missing, rather than merely that the account number had been omitted from the first card of the job.

Although the improved messages in this experiment were longer than either the UNIVAC or IBM messages, it is not clear that longer is always preferable. The terminology, phrasing, and content seem far more important than the mere length of the message. It seems that systems designers have not always been sensitive to the capabilities of novice users and that there is room for improvement. Proposed messages should be tested for comprehensibility with several levels of users.

3. CONCLUSIONS

This series of five studies on system messages in COBOL-compiler, text-editor, and job-control language processor applications supports the conjecture that the wording and content of messages can impact novice user performance and attitude. Rigid guidelines cannot be made since the subtleties of natural language usage are hard to formalize or quantify. However, the arguments in Section 1 and the results of Section 2 provide some guidance to system designers who are sensitive and eager to serve the user community. The studies described in Section 2 should also serve as a model for design studies which could be carried out in the product development process. The intuition of the designer can be supplemented by simple, fast, and inexpensive design studies with actual users and several alternative messages.

If the project goal is to serve novice users, then ample effort must be dedicated to designing, testing, and implementing the user interface. This commitment must extend to the earliest design stages so that programming-language, command-language, or menu-selection approaches can be modified in a way which contributes to the production of specific error messages. Messages should be evaluated by several people and tested with suitable subjects. Messages should appear in user manuals and be given greater visibility. Records should be kept on the frequency of occurrence of each error. Frequent errors should lead to software modifications which provide better error handling, to improved training, and to revisions in user manuals.

I have often felt that users are more likely to remember the one time when they had difficulties with a computer system rather than the 20 times that everything went well. The strong reaction to problems in using computer systems comes in part from the anxiety and lack of knowledge that novice users have. This reaction may be exacerbated by a poorly designed excessively complex system, a poor manual or training experience, or from the hostile, vague, and irritating system messages that they receive. Improving the messages will not turn a bad system into a good one, but it can play a significant role in improving the user's performance and attitude.

4. ACKNOWLEDGMENTS

Control Data Corporation provided partial support for the preparation of this report and the University of Maryland Computer Science Center provided text processing support.

This research could not have been done without the enthusiastic efforts of my students who carried out the experiments described in Section 2:

Section 2.1 Patrick Peck and David Fuselier
Section 2.2 Tony Jefferson and Mae Chang
Section 2.3 Richard M. Dudley, Nancy Hale and Joel Buxton
Section 2.4 Sharon Nelson
Section 2.5 Mark S. Fisher and C. Patrick Zilliacus

I am also grateful for the numerous comments from Barry Dwyer, John Gannon, William Mosteller, James Reggia, Elliot Soloway, Bonnie Webber, Mark Weiser, and Marvin Zelkowitz.

5. REFERENCES

Conway, R.W., & Maxwell, W.L. CORC the Cornell computing language. *Communications of the ACM,* 1963 *6* (6), 317–324.

Conway, R.W., & Wilcox, T.R. Design and implementation of a diagnostic compiler for PL/I. *Communications of the ACM,* 1973, *16* (3), 169–179.

Cress, P., Dirksen, P, & Graham, J.W. *Fortran IV with WATFOR and WATFIV*. Englewood Cliffs, NJ: Prentice-Hall, Inc., 1970.

Dwyer, B. Programming for users: A bit of psychology. *Computers and People*, 1981, *30* (1 and 2), 11–14, 26.

Dwyer, B. A user friendly algorithm. *Communications of the ACM*, September 1981, *24* (9), 556–561.

Golden, D. A plea for friendly software. *Software Engineering Notes*, 1980, *5* (4).

Hahn, K.W., & Athey, J.G. Diagnostic messages. *Software—Practice and Experience*, 1972, *2*, 347–352.

Heaps, H.S., & Radhakrishnan, T. Compaction of diagnostic messages for compilers. *Software—Practice and Experience*, 1977, *7*, 139–144.

Mosteller, W. Job entry control language errors. *Proceedings of SHARE 57*. Chicago, IL: SHARE, Inc., 1981, 149–155.

Moulton, P.G., & Muller, M.E. DITRAN—a compiler emphasizing diagnostics. *Communications of the ACM*, 1967, *10*, 45–52.

Norman, D.A. The trouble with UNIX. *Datamation*, November 1981, *27* (12), 139–150.

Roberts, T. *Evaluation of computer text editors*. Xerox Palo Alto Research Center Report SSL-79-9, November 1979.

Woody, C.A., Fitzgerald, M.P., Scott, F.J., & Power, D.L. A subject-content-retriever-for-processing-information-oning- (SCORPIO). *AFIPS Conference Proceedings*, 1977, *46*.

Zelkowitz, M.V. *PL/1 Programming with PLUM*. Geneva, IL: Paladin House, 1976.

4

Artificial Intelligence and Human Factors Engineering: A Necessary Synergism in the Interface of the Future

PAUL ROLLER MICHAELIS, MARK L. MILLER
AND JAMES A. HENDLER

Computer Science Laboratory
Central Research Laboratories
Texas Instruments Incorporated
Dallas, Texas 75266

In the coming decade, a new generation of computer-based systems offers the potential to be an amplifier and catalyst for human thought. The authors are members of a team that is attempting to construct and evaluate experimental prototypes of such systems. This team combines talent from many diverse disciplines, among them artificial intelligence and human factors engineering. The specialists in these two disciplines have developed a synergistic relationship that is illustrated in this chapter, using as examples two ongoing research projects.

One project addresses a major problem with existing interactive natural language computer systems: they are often unable to cope with the complexities of unrestricted natural language dialogue. One solution is to improve the quality of the natural language processors. This chapter decribes the initial efforts in a different approach, in which artificial intelligence and human factors specialists are developing a computer-processable, human-engineered subset of natural language. The second research project is in the area of advanced computer based instruction. The goal is to develop a programming tutor that is able to diagnose a student's conceptual problems and provide the student with custom-generated help.

1. INTRODUCTION

People will have trouble performing a physical task, if the demands of the task exceed their physical capacities. To many of us nowadays, that seems like simple common sense. However, it was not until the late 1890s that Frederick W. Taylor made his pioneering studies of how to design jobs and tools so that they more closely match the physical capabilities of people.

The field of human factors engineering had its birth during World War II. The founders of the field recognized something that today is also considered common sense: errors can occur in human-machine systems when the person's job in these systems overloads his or her *mental* capacities. Researchers such as Alphonse Chapanis and Paul Fitts discovered one factor in particular that greatly contributed to mental overload: errors were much more common in systems where the format for information exchange between person and machine conflicted with the way the person thought about the information. For example, when analyzing airplane accidents that had been attributed to "pilot error," they found that certain control and display designs virtually invited even experienced pilots to misuse or misinterpret them.

The solution lay in making the operation of human-machine interfaces compatible with the mental processes and capabilities of humans. This remains a chief goal of human factors specialists, especially as new computer technologies add to the complexity of human-machine systems. One way to enhance human-computer interaction is to develop computer software that is "intelligent."

Efforts to create artificially intelligent interactive systems have been underway for only a couple of decades. Most of the researchers who are credited with founding the field of artificial intelligence are still active contributors (see, e.g., McCarthy, 1965; Minsky, 1966; Newell & Simon, 1972). Systems have been developed that can play master-level chess, solve complex integrals, understand and obey commands stated in simple English, speak in a human-like voice, recognize objects in scenes, solve analogy problems, and so on.

An immediate goal of this work is to develop software that will free humans from having to do many of the monotonous, low-level tasks they have typically been assigned in human-machine systems. A long-term goal is to develop interactive systems that can serve as an amplifier and catalyst for human thought. To be successful, these artificially intelligent systems will need to exchange information smoothly and efficiently with their human operators. Therefore, the teams of people who design these systems need to contain both aritificial intelligence and human factors specialists. The authors are members of such a team. In describing two research projects supervised by the authors, this paper illustrates how artificial intelligence and human factors specialists can work together to design intelligent interactive systems.

2. INTERACTIVE NATURAL LANGUAGE SYSTEMS

What is the best way to interact with a computer in order to solve problems? Three alternatives are generally offered: menu selection, query languages, and natural language. From a human factors standpoint, in many cases menu selection is an excellent means of interaction. It also has the added advantage of placing a minimal parsing burden on the computer. However, for certain less-structured applications, such as conversational interaction, menu systems are inadequate because they severely limit the strategies available to the user. Query languages place a greater parsing burden on the computer, and are somewhat less limiting to the user. However, they suffer from human factors problems in that they are often very difficult for the user to learn (Reisner, 1981). Unrestricted natural language interaction is the least limiting to the user, but it is simply not machine-parsable using present techniques.

Chapanis (1975) has demonstrated that interactive natural language dialog is remarkably unruly, with many misspellings and grammatical errors. Although progress has been made in getting computers to process pristine English text, it will be many years before computers will be able to process unlimited interactive natural-language dialog.

2.1 Description Of The Problem

As researchers work toward a system that interacts in true natural language, another project is under way that is oriented toward intermediate results. The goal of this project is to define a human-engineered subset of natural language that would place a much smaller processing burden on the computer. The key to this language subset's success is that it will retain nearly all of the user-oriented bene-fits claimed for unrestricted natural language dialog. Specifically, this language subset should be easily and quickly learned even by computer-naive users, easy to use and highly versatile once it is learned, neither intimidating nor frustrating for the user, and not easily forgotten even after long periods of time. Further, when compared with unrestricted natural language, the use of this language subset in the performance of a task should not adversely affect the time required or the strate-gies employed by users to accomplish that task.

2.2 What Human Factors Engineering Contributes

The authors of this paper are not by any means the first to propose restricting users' dialog to specific subsets of English. Existing systems that require users to structure their dialog in particular ways include Kaplan's CO-OP (1979), and Biermann and Ballard's NLC (1980). In the former, the user is limited to only asking questions beginning with ''WH'' words, and in the latter the user is re-stricted to giving commands that start with an imperative verb. Each of these limi-tations greatly reduces the syntactic coverage required and also precludes many other types of utterance such as comments by the user on the course of the dialog, context setting, and attempts to define new concepts.

Although successful for certain narrow domains, for many other domains this type of dialog restriction imposes an unacceptable limit on what can be ex-pressed by the user. The research reported below, therefore, takes a different ap-proach: one intended to discover language restrictions that make users' inputs more easily machine understandable without degrading users' ability to express concepts.

Ford, Weeks, and Chapanis (1980) and Michaelis (1980) reported a series of experiments that were conducted in the human factors laboratory at The Johns Hopkins University. In these experiments, two-person teams exchanged information over a telecommunications medium in order to solve problems. Half of the teams in each experiment were rewarded solely for correctly solving their problems. The other half had their correct-solution reward diminished for each word token they used. Thus, these latter teams were encouraged to keep their communication as brief and concise as possible. An example of a protocol from an unlimited team in the Michaelis study is shown in Appendix A, and a protocol from a limited team appears in Appendix B. The problem-solving task assigned to the subjects in the Michaelis experiment is typical of the type used in these studies: One team member was given a completely assembled prism-shaped wooden model, and was required to assist the other member, who had to build an identical model from the separate parts. In these experiments, the team members were in different rooms. In the Ford et al. study, half the teams communicated by voice and the other half via teletypewriters; in the Michaelis study, all communication was over teletypewriters.

In both studies, there were dramatic and highly significant differences between the two experimental groups. However, it is important to note that problem-solving accuracy was not affected by self-imposed brevity.

Among the significant differences noted in both studies are that the self-limited teams generated, on the average, about one fifth as many word tokens, one third as many word types, and one third as many messages. In a linguistic analysis of the protocols from their study, Ford et al. found that the self-limited subjects used proportionally more nouns (41.9 vs. 26.1%, $p < .001$), fewer pronouns (5.5 vs. 11.9%, $p < .001$), fewer verbs (10.3 vs. 16.9%, $p < .001$), more adjectives (18.3 vs. 10.4%, $p < .001$), and fewer prepositions (8.9 vs. 11.3%, $p < .035$).

Probably the most interesting finding of these studies is that, on the average, the self-limited teams solved their problems faster than their unlimited counterparts, 14.9 vs. 19.3 minutes in the Ford et al. study and 20.5 vs. 27.6 minutes in the Michaelis study. This difference was not statistically significant in the Ford et al. study. However, in the Michaelis study, which tested more teams (48 vs. 32), the p value was less than 0.005. This is strong evidence that requiring people to be concise does not hurt their ability to communicate; it may even help.

2.3 What Artificial Intelligence Contributes

After these original studies were conducted, specialists in artificial intelligence became involved. They contrasted the limited and unlimited protocols from the Michaelis study. Their goal was to determine how the dialog limitation might affect the processing burden of natural language computer systems. They were specifically concerned with contrasting the effects on systems that do a syntactic analysis first and then pass the results to a semantic component, versus those which integrate the semantic and syntactic components during analysis.

Among their findings are that pronominal reference and the attachment of prepositional phrases, two stumbling blocks for many present syntactically based systems, occur somewhat less frequently in the limited condition. However, in the limited protocols, over one third of the utterances were ungrammatical, while in the unlimited case, this was closer to one tenth. They therefore believe that syntax-first approaches will have significantly more problems parsing the limited condition utterances than systems which have less reliance on syntax.

On the other hand, the word types used in the limited condition are virutally a subset of those used by the unlimited users; apparently, many of the words used by the unlimited subjects were not necessary for the solution of the problem. This finding has also been reported in a study of interactive limited-vocabulary dialog (Michaelis, Chapanis, Weeks, & Kelly, 1977), and suggests that the conceptual coverage of the limited protocols is less than that of the unlimited. Therefore, a system that relies on semantics, such as a semantic grammar (c.f. Burton, 1976) or conceptual analyzer (c.f. Schank, 1975), could possibly gain efficiency from this language limitation.

The protocols were also analyzed to determine whether the problem-solving strategies used were different between the unlimited and limited conditions. The protocols were classified according to the problem-solving strategies used and the ordering of their subgoals. No statistically significant differences were found between the unlimited and limited conditions in the number of teams using the different strategies.

In 38 of the 48 protocols (19 in each condition), the subjects used subgoals characteristic of classic means-ends analyses (Newell & Simon, 1972). These teams established two major subgoals of the task, building the triangular sides and the rectangular base. The order in which these were performed did not significantly differ between the limited and unlimited conditions.

The 10 remaining teams did not have obvious subgoals; 6 used an approach in which they described the appearance of the model, and the remaining 4 used a strategy of making small pieces and then connecting these together. Again, no significant differences were found between the two conditions in the number of teams using each strategy.

To summarize the findings thus far in this research effort, human factors specialists found no evidence that the dialog restriction discussed in this paper will hurt users' efficiency. Indeed, the Michaelis study suggests that the efficiency of the users may actually be improved by well-chosen limitations on the interactions. Further, the language restriction could not be shown to significantly change the problem-solving strategies used by the subjects. The protocol analyses performed by artificial intelligence specialists suggest that semantically based interactive natural language processing systems might also benefit from this restriction.

2.4 Work In Progress

As was stated earlier, artificial intelligence specialists have developed many techniques intended to allow natural language "understanding" by computers. None of these techniques is able to process unrestricted interactive natural language dialog. The analysis of concise dialog was the authors' first step toward developing a human-engineered language subset that can be processed by existing natural language techniques. Although user conciseness might reduce the computer's processing burden, this restriction on dialog is not by itself sufficient. If existing natural language processing techniques are to be successfully used in interactive systems, the dialog will need to be more limited in its syntax and semantics.

Regarding syntax, Malhotra (1975) has reported that 78% of the utterances generated by subjects could be classified into 10 simple syntactic types. In examining subjects using the PLANES system, Tennant (1980) found that the syntactic types were somewhat different, but that most sentences could be described by a few simple syntactic classifications.

The authors are presently conducting studies to discover whether users can adapt to a subset of English grammar consisting entirely of common syntactic structures similar to those reported by Malhotra and Tennant. The experimental methodology is based on that developed by Kelly and Chapanis (1977) in their study of interactive limited vocabulary dialog. Kelly and Chapanis first determined what words were used by unrestricted two-person teams to solve particular problems. In a subsequent experiment, they required two-person teams to solve identical problems using only the 300 words most commonly used by the unrestricted teams. In this experiment, the restricted teams solved their problems as quickly and accurately as the teams in the unrestricted control group. (Word usage by the Kelly and Chapanis subjects is described by Michaelis, Chapanis, Weeks, & Kelly, 1977.) It is the authors' hope that, just as subjects can quickly adjust to using only words from a well-chosen limited vocabulary, they may also be able to restrict themselves to a well-chosen set of syntactic constructions.

Even if researchers develop an ideal set of syntactic and vocabulary limitations, there still remain problems with semantic limitations. In particular, natural language systems suffer from what has been referred to as semantic overshoot (Codd, Arnold, Cadiou, Chang, & Roussopoulos, 1978). This is an attempt by the user to exceed the conceptual coverage of a natural language processor. For example, in a suppliers-projects-and-parts domain, a reference to a manager would be an instance of semantic overshoot if there is no data on managers in the data base. Since present natural language systems have limited conceptual coverage, semantic overshoot is a common problem.

It is important to try to identify the occurrence of semantic overshoot so that an appropriate error message can be reported to the user. If the problem is not identified and the user merely gets a content-free response like, "Please rephrase your question," he has no clue as to whether the problem was a spelling error, poor syntax, an unknown word, or semantic overshoot. Indeed, even if, in the above case, the system responded, "I don't know the word MANAGER," it is still not clear to the user whether he has used an illegal word or introduced an illegal concept. Codd et al. suggested that the system maintain a set of keywords related to but not part of the domain of discussion. This would allow the system to identify instances of semantic overshoot, in this case by recognizing the word MANAGER. The system could then inform the user that he had exceeded the conceptual coverage of the system.

Although it is certainly a step in the right direction, the problem of semantic overshoot is not solved by simply detecting and reporting overshoots to the user. Systems that operate in this manner may deny users the freedom to develop and employ their own strategies, thereby negating a major advantage of natural language interaction. Ideally, natural language systems should be able to accept all reasonable user strategies. Therefore, along with developing human-engineered syntactic and lexical subsets, the authors are using similar methodologies to develop human-engineered semantic subsets. Experiments, like those described in Section 2.2, are conducted to obtain protocols of two people communicating to solve problems. Then, as is described in Section 2.3, these protocols are analyzed to find the frequency with which different strategies are employed. Semantic overshoot should be less likely in systems that include in their semantic coverage these common user strategies. Although such systems will still have limited semantic coverage, in theory the limitation will be rarely noticed by users. Combined with the experimentally determined syntactic and lexical limitations, the authors expect the result to be a computer-processable subset of natural language that does not hamper people's ability to communicate.

3. INTELLIGENT TUTORING SYSTEMS

The previous section described the efforts of artificial intelligence and human factors specialists to develop a computer-processable, human-engineered subset of natural language. Another area in which artificial intelligence and human factors specialists are cooperating is in the development of "intelligent tutoring systems" intended to teach elementary computer programming. Such systems represent enhancements over conventional "drill and practice" or "frame-based" multiple-choice branching systems, because they incorporate considerable knowledge about the task, the student, and tutoring per se. The long-term goal is to provide a computer-based educational experience comparable to a one-on-one interaction with an expert human tutor.

3.1 Description Of The Problem

Three systems intended to teach elementary computer programming are examined. The first system, the "BASIC Instructional Program," or BIP (Barr, Beard, & Atkinson, 1976), serves as a problem-solving laboratory wherein students solve programming exercises in the BASIC language. The second system, Miller's (1979) "Structured Planning and Debugging Environment," SPADE-0, was developed to articulate the underlying programming knowledge involved in writing graphics programs in the LOGO language (Papert, 1980). A third system, currently under development, blends both artificial intelligence and human factors considerations, and is intended to teach a LOGO-like language to computer-naive students.

BIP is an effective working computer-based instructional system which has been in use by the Navy for some time, both as a research vehicle and as a practical tool for teaching BASIC programming. It is a good example of previous work on programming tutors, and has been in use in the authors' laboratory as a bench mark since early 1980. BIP is able to individualize the sequence of tasks encountered by a given student, based on a model of the student's skills it infers from that student's performance on prior tasks. Hence, not every student encounters every task, and some encounter tasks in a different order than others. Below is a slightly edited excerpt from a session with this system that illustrates some of its capabilities and limitations.

```
COMPUTER: On the first day of XMAS, someone's true love
          sent him a partridge in a pear tree. On the
          second day, they sent two turtle doves plus
          another partridge (3 gifts, 2nd day). This
          continued through the twelfth day, when the
          true love sent 12 lords, 11 ladies, . . . all
          the way to yet another partridge. Write a
          program that computes and prints the number
          of gifts sent on that twelfth day. (This is,
          not the same as the TOTAL number of gifts
          sent for all 12 days!)

 STUDENT: HINT?

COMPUTER: This program requires a loop. Each execution
          of the loop involves accumulating the value
          of the index into a total.

 STUDENT: HINT?

COMPUTER: Finding a total almost always means you need
          two things:
          (1) Setting a variable to 0 before the loop.
          (2) Accumulating values into that variable
          within the loop.
```

BIP has impressive features for creating flow diagrams and does an excellent job of sequencing tasks. However, its understanding of the domain is limited to a flat collection of language constructs. Based on informal analyses, BIP often rejects answers that students believe to be correct. It insists on eliciting a single solution, and this lack of flexibility is not always appropriate in complex domains such as programming.

BIP was hampered by its lack of understanding of planning and debugging—two central concerns when teaching programming. While BIP could individualize the sequence of tasks, it could not individualize the hints given within a task. Thus, all students who encountered the XMAS task and requested two hints would see the same two hints shown above. To improve upon BIP's pre-stored hints, the problem faced by the authors was twofold: to represent the underlying knowledge and to apply that knowledge in a fashion helpful to the human user.

3.2 What Human Factors Engineering Contributes

Simply stated, the goal of this project is to design computer environments that tutor students in much the same way that a human teacher might tutor his or her students. In order to accomplish this goal, some very basic questions must be answered: What makes intelligent human tutors successful? What are their techniques for diagnosing student problems and misconceptions? What are their techniques for advising students? In short, how do they use their intelligence to provide tutoring superior to that provided by pre-stored hint systems like BIP?

Human factors engineers addressed these questions by setting up a system in which a computerized intelligent tutor is simulated by having an intelligent human playing the role of the computer tutor. Very simply, the human tutor observes a student's efforts by watching a monitor that is slaved to the student's work terminal. The tutor makes judgments about the student's problems and misconceptions, and types appropriate help messages that appear on the student's help terminal. It is important to recognize that, in this paradigm, the human tutor bases decisions on exactly the same information that would be available to the computer tutor, and similarly provides help the same way that the computer tutor should.

In these studies, the human tutor is carefully evaluated. His or her activities are meticulously recorded, along with verbal protocols in which the tutor explains the rationale behind his or her decisions. These studies are not yet complete, but a clearer model of the intelligent human tutor is already emerging. One trend observed thus far is that many of the diagnostic processes of the human tutor could probably be performed by a relatively simple computer tutor.

Here is an example of a problem a student had that was easily diagnosed by the human tutor. The student was learning how to program in BASIC, using the BIP problem set. In this particular problem, the student was asked to take two numbers, M and N, and compute their sum, difference, product, and quotient. This is what the student typed:

```
10 PRINT ""WHAT IS THE FIRST NUMBER""
20 INPUT M
30 PRINT ""WHAT IS THE SECOND NUMBER""
40 INPUT N
50 LET A = M + N
60 LET B = M - N
70 LET C = M * N
80 LET D = M
```

At this point, the student paused for over a minute, then asked for help. Quite clearly, the student's problem was that he did not know the symbol for division. This sort of problem is representative of the type solved by the human tutor that would not have been solved by a pre-stored hint tutor like BIP. Note that even a very simple means-ends analysis model involving sequential accomplishment of subgoals is adequate to provide a correct hint here.

On the other hand, in some cases even highly skilled human tutors were unable to diagnose student problems solely by examining their programming efforts. It therefore appears likely that, when teaching programming, the capability for some form of student-tutor dialog is desirable.

3.3 What Artificial Intelligence Contributes

The crucial contributions of artificial intelligence to computer-assisted instruction involve representing the underlying knowledge. In the case of programming, representing the domain knowledge requires asking such questions as, "What is it that the expert programmer knows that the novice does not?" Miller's SPADE-0 project was more an attempt to investigate and formalize this type of knowledge than to build a useful programming tutor. It represented knowledge about programming plans (i.e., procedural templates independent of the particular programming language) and debugging techniques.

SPADE-0 built upon work in automatic planning and debugging developed in HACKER (Sussman, 1973), MYCROFT (Goldstein, 1974), and NOAH (Sacerdoti, 1975). SPADE-0 could prompt the student through hierarchical planning processes, encouraging the student to postpone premature commitment to the detailed form of the code. This system provided a vocabulary of concepts for describing plans, bugs, and debugging techniques, and handled the routine bookkeeping tasks involved in simple program development.

The key feature of SPADE-0 is its deeper analysis of the student's underlying knowledge. This is manifested by commands for editing the plan—rather than merely the code—of the student's program. However, the design of SPADE-0 ignored human-factors considerations, imposing its own technical vocabulary on the student, and adopting a style of interaction that took away much of the initiative.

3.4 Work In Progress

The authors' current work is an attempt to extend the underlying knowledge represented by SPADE-0, and package it according to human factors guidelines. Like BIP, the new tutor will dynamically select tasks from a curriculum data base; but like SPADE-0, it will build a model of the student's problem-solving skills, rather than simply record which programming language constructs have been mastered. The key contribution of artificial intelligence is the ability to do a fine-grained diagnosis of student errors and provide custom-generated (rather than pre-stored) advice.

The design of the new tutoring module is being based on the human factors studies in which a human simulates the ideal module. As the system implementation progresses, additional tasks will be taken over by the computer, and the need for the human tutor to intervene will be correspondingly diminished. The proportion of tasks successfully performed by the computer tutor is a measure of progress.

Earlier "intelligent tutoring systems" such as BIP and SPADE-0 built models of the student. However, the interface between the tutor and the student remained crude. By working with human factors engineers, the artificial intelligence specialists now better understand how *human* tutors interact with students. The emphasis of the work has now shifted to modeling this tutor/student interface.

4. CONCLUSION

In closing, it is worthwhile to review a central human factors problem: the division of labor between human and machine in human-machine systems. In any well-designed system, tasks are allocated to those components best suited to perform them. Textbooks on human factors engineering typically state that machines tend to be superior to humans in such tasks as calculation and coordination of many simultaneous activities. Conversely, they state that humans tend to excel in such tasks as problem solving where originality is required, pattern recognition, and decision making based on incomplete or conflicting data, or when unlikely or unexpected events occur. Thus, these guidelines would allocate responsibility for calculation to the machine, but leave the human responsible for recognizing and basing decisions upon patterns in the results of those calculations.

As artificial intelligence continues to progress, machines will be able to successfully perform more and more tasks that have traditionally been assigned to humans. This might lead to speculation that research on human-machine interfaces may be unnecessary, since the need for the human component will disappear. For certain kinds of menial tasks presently performed by humans, this line of reasoning is probably sound. However, it is the authors' expectation that, as work in artificial intelligence and human factors engineering continues to advance, the nature and power of the human-computer interface will become *more* critical and sophisticated. The art and science of interface design will never become obsolete. Obsolescence is faced only by the traditional human-machine system task-allocation guidelines.

This paper has described two research projects in which artificial intelligence and human factors specialists have collaborated. From these projects and others like them, the authors have learned to stop thinking in terms of separate disciplines that merely benefit from cooperation. Particularly in the design of "intelligent interactive systems," the borderline between these two fields has blurred. Human factors specialists are learning to exploit the tremendous benefits for people made possible by more intelligent software components; artificial intelligence specialists are learning to write software that is sensitive to the needs, capacities, and limitations of human beings. It is the authors' expectation that, due in large part to this synergism, the human-computer interface can become an extraordinary catalyst for human thought.

5. ACKNOWLEDGMENTS

The research on tutoring systems was funded in part by the Office of Naval Research, Psychological Sciences Division, Contract Number N00014-80-C-0818.

The authors wish to thank Dr. Harry R. Tennant for his helpful comments and additions to this manuscript.

6. REFERENCES

Barr, A., Beard, M., & Atkinson, R. The computer as a tutorial laboratory: the Stanford BIP project. *International Journal of Man-Machine Studies,* 1976, *8,* 567–596.

Biermann, A.W., & Ballard, B.W. Toward natural language computation. *American Journal of Computational Linguistics,* 1980, *6* (2), 71–86.

Burton, R.R. *Semantic grammar: An engineering technique for constructing natural language understanding systems* (BBN Report 3453). Boston, MA: Bolt Beranek & Newman Inc., 1976.

Chapanis, A. Interactive human communication. *Scientific American,* 1975, *232* (3), 36–42.

Codd, E.F., Arnold, R.S., Cadiou, J.M., Chang, C.L., & Roussopoulos, N. *Rendezvous version 1: An experimental English-language query formulation system for casual users* (Report RJ2144 [29407]). San Jose, CA: IBM Research Laboratory, 1978.

Ford, W.R., Weeks, G.D., & Chapanis, A. The effect of self-imposed brevity on the structure of diadic communication. *The Journal of Psychology,* 1980, *104,* 87–103.

Goldstein, I. *Understanding simple picture programs* (Artificial Intelligence Laboratory Report 294). Cambridge, MA: Massachusetts Institute of Technology, 1974.

Kaplan, S.J. *Cooperative responses from a portable natural language data base query system* (Report HPP-79-19). Stanford, CA: Stanford University, Heuristic Programming Project, 1979.

Kelly, M.J., & Chapanis, A. Limited vocabulary natural language dialogue. *International Journal of Man-Machine Studies,* 1977, *9,* 479–501.

Malhotra, A. *Design criteria for a knowledge-based English language system for management: An experimental analysis* (Project MAC Report TR-146). Cambridge, MA: Massachusetts Institute of Technology, 1975.

McCarthy, J. *LISP 1.5 programmer's manual.* Cambridge, MA: MIT Press, 1965.

Michaelis, P.R. Cooperative problem solving by like- and mixed-sex teams in a teletypewriter mode with unlimited, self-limited, introduced and anonymous conditions. *JSAS Catalog of Selected Documents in Psychology,* 1980, *10,* 35–36 (Ms. No. 2066).

Michaelis, P.R., Chapanis, A., Weeks, G.D., & Kelly, M.J. Word usage in interactive dialog with restricted and unrestricted vocabularies. *IEEE Transactions on Professional Communication,* 1977, *PC-20,* 214–221.

Miller, M.L. A structured planning and debugging environment for elementary programming. *International Journal of Man-Machine Studies,* January 1979.

Minsky, M. Matter, mind, and models. *Proceedings of International Federation of Information Processing,* 1966.

Newell, A., & Simon, H.A. *Human problem solving.* Englewood Cliffs, NJ: Prentice-Hall, 1972.

Papert, S. *Mindstorms.* New York: Basic Books, 1980.

Reisner, P. Human factors studies of database query languages: A survey and assessment. *ACM Computing Surveys,* 1981, *13* (1), 13–32.

Sacerdoti, E. *A structure for plans and behavior.* New York: Elsevier, 1975.

Schank, R.C. *Conceptual information processing.* New York: Elsevier, 1975.

Sussman, G. *A computational model of skill acquisition* (Artificial Intelligence Laboratory Report 297). Cambridge, MA: Massachusetts Institute of Technology, August 1973.

Tennant, H.R. *Evaluation of natural language processors* (Report T-103). Urbana, IL: University of Illinois, Coordinated Science Laboratory, November 1980.

7. APPENDIX A

Below is a full protocol from one of the unlimited teams in the Michaelis (1980) study. SK is the Seeker, the team member who had to assemble the model using information provided by SO, the Source. Although the margins have been straightened, all of the spelling and grammatical errors of the original protocol have been preserved in this transcription. The reader is reminded that this protocol is an example of true unrestricted natural-language dialogue.

```
SK:  What does the model look like?

SO:  What I'll try to do is put /tha/ the model together
     with you. First there are three /f/ round pieces/s/
     with holes in them, /a/ that are put in a triangle
```

shape, by attaching two yellow sticks to one of the pieces,, taking the other two pieces and, these are the other round parts and attaching them to the ends of the yellow sticks, and then taking a blue stick to close the triangle. Question?

SK: I have only two Never mind I just realized what to do

SO: Do you have this part of the model so far?

SK: Not yey

SO: To repeat, three round pieces in a triangle. Two yellow stickes coming out of one at a ninety degree angle. Attached to the yellow sticks, the other two round pieces. Now connect these two round pieces with a blue stick to make a triangle (isosceles). All right? When the yellow sticks are in the one round piece there is one hole separating them. When the blue piece is in the two round pieces there are no holes separating the yellow stick and the blue stick. etter? The three round /i/ pieces are all the same. They are not the pieces with the holes (4) going through their depth.

SK: Okay I have figured it out, what next?

SO: Now take the two round pieces with the holes through them, and slide then on, each on onto a green stick that is attached to the round pieces which are connected to the blue stick. Actually, to make this easier, the two round pieces iwht the holes are connected by another blue stick. It would probably be easier to first connect them and then slide them on to the green sticks.

SK: Do I connect the two rounde pieces through the center hole or throught the sides?

SO: The sides.

SK: And wh how do I put them on the green sticks

SO: Through their center holes. Can I go on?

SK: Yes

SO: Now make the same triangle structure we did in the very /e/ beginning. This will attach the same way as the first one did, and it will close the structure. Start to do this; the end model shoud look like the roof of a house, where the ''attic'' has a sliding piece along it. If you have this, we're finished.

SK: I have two extra pieces, do I need them for anything?

SO: I'll count the number of pieces or my model and tell you how many there are total needed. There are eighteen different pieces used. Do you have too many?

```
SK:  I have only fifteen pieces with an extra blue stick
     and two strang rounded pieces

SO:  Only three blue pieces should be in the model. Is
     that how many you have in the model?

SK:  No ihave only two

SO:  Okay, once again. Each triangle, have two yellow
     sticks and one blue stick. Do you have that?

SK:  Yes

SO:  There is a sliding piece made up of two round pieces,
     which are different /d/ from the rest because they
     have four holes going through their cent/t/er besides
     the center hole. They are connected by a blue stick.
     Do you have that?

SK:  No

SO:  Remember I talkied about the sliding piece. This is it.

SK:  Do I connect the two rounded pieces

SO:  Remember, you connect them through their sides, and
     then slide the structure onto two green sticks through
     their centter holes. Do you have that?

SK:  no

SO:  What's the problem that I can help you with?

SK:  I have the roof part but I dont know wher the
     attic-slidding piece goes

SO:  If you have the roof all constructed, take off one
     triangle side. Now, take the sliding structure and
     slide it onto the two green sticks through the center
     holes of the weird round pieces, so that all the blue
     sticks are parallel. Now do you have that? Then
     connect the triangle back on to/p/ make a full m

SK:  Okay I have finished it I will now disconnect the
     machine, okay?

SO:  Fine.
```

8. APPENDIX B

This is a full protocol from one of the self-limited teams in the Michaelis (1980) study.

```
SK:  begin

SO:  place 3 discs containing one hole passing through the
     center on the table to form a triangle. In one of
     these place two yellow pegs in
```

SK: the pegs go where in the disc

SO: in the holes on the side so there is one hole between
them

SK: continue

SO: place one disc at the end each peg, connect the tht
new discs on the side holes, then place blue peg
between new discs. all discs same plane. Construct
identical model.

SK: okay

SO: disc with five holes through center, place on table,
connect to identical disc using blue peg. Discs in
same plane.

SK: oks

SO: in original construction, place one green peg in each
center hole.

SK: yes

SO: paired discs fit loosely over green pegs, move freely

SK: okay,

SO: place remaining construct on green pegs at other end
from original. check: all blue pegs in a plane,

SK: yes is it finished?

SO: yes

5

Psychological Investigations of Natural Terminology for Command & Query Languages

SUSAN T. DUMAIS
THOMAS K. LANDAUER
Bell Laboratories
Murray Hill, NJ

It is frequently asserted that novice and unsophisticated computer users would find computer systems more congenial and easier to use if communications with them employed more "natural" words. In a series of empirical studies, we have examined some characteristics of natural command and query terminology, and tested the value of a natural command lexicon in the initial stages of learning a computerized text-editing system. Our conclusion is that while available methods can reveal important and interesting properties of naive users' conceptions of potentially computerized tasks, the complicated interaction between humans and computer systems is still in need of much more systematic study before generally valid precepts for the design of interactive languages are available.

1. INTRODUCTION

It is often asserted by system designers and psychologists alike that naive users would find computer systems more congenial if the communication protocol and lexicon were more "natural." What we mean by "natural" here is something that naive and casual users spontaneously do. (But see also, Jackson & Tschergi, 1981 for comments about the difficulties in defining a natural command language.) One idea about how to make a system easy for novices to use is to look at how people now perform similar non-computerized tasks, and incorporate (and perhaps embellish) the habits people have developed in such environments, in the computer environment.

Information retrieval is an example of a task that was once done primarily manually, but which is now being computerized. People often attempt to retrieve specific information from computerized sources— e.g., bibliographic reference and airline reservation systems. For the most part, these tasks are still performed by specialists. More and more, however, casual users use these or similar information retrieval systems (e.g., office automation and home information systems).

Several methods can be used to retrieve information from large data bases. Four examples are: menu selection; keyword search; special query languages; and "natural" language specification. Without going into any great detail, we can note that each of these retrieval schemes has some problems. (For a more detailed discussion of experimental results, see Allen, this volume.)

1. *Menu-selection schemes* are employed in many systems designed for nonspecialist users. The main advantage of a menu-selection technique is that no special vocabulary need be learned by the user; specification depends on recognition rather than production. Another advantage is that menus structure the choice space. Thus, users are not supposed to require extensive special training. There are problems, however: the word meanings and categorization schemes invented by system designers may not match those of the end user; the organization imposed by menu-selection schemes is usually quite inflexible, providing only a single access route to a given item (but see Furnas, 1981); and, in large data bases, menu selection can be tedious.

2. Retrieval in *keyword systems* is typically based on user provided single words or Boolean combinations of keywords. Several psychological investigations have shown that Boolean expressions (Ackroff, 1979) and quantifiers (Thomas, 1977; Thomas & Gould, 1975) are difficult for people to formulate or understand, even after extensive training. For a wide variety of tasks, natural logic seems to differ significantly from formal logic (Braine, 1978). Furthermore, knowledge of the cataloging or indexing system itself (in addition to the content area of interest) can be necessary to the selection of the appropriate keywords (Stevens & Shneiderman, 1981).

3. *Special data-base query languages* (such as, SQUARE, SEQUEL, QBE, and RAMIS) can be very powerful for well-trained users. But they encounter the usability problems of both menu and keyword systems. That is, they often have predefined categories that may not match those of the user, and they require formulation of difficult logical search expressions. Furthermore, general query systems usually require user's to have a detailed knowledge of the data and data organization. (See Reisner, 1981, for a more comprehensive review of psychological issues in data-base query languages.)

4. *Natural-language* based systems are another possibility. Many have argued that since people already know their natural language, they would not have to invest time in learning special purpose data-base query or programming languages. But, people actually have great difficulty in formulating precise and unambiguous requests in English (Shneiderman, 1981). Furthermore, natural-language systems involve tremendous overhead to develop, and are currently restricted to very limited domains, even when they can be implemented. And, finally, natural language systems must sometimes overcome unrealistic expectations of the computer's power, including requests for information not contained in the data base, and questions involving value judgements.

2. STUDY 1

We thought that some interesting alternatives or modifications to existing systems (that might make them easier for novices to use) would be suggested by looking at how people now obtain information from non-computerized sources (and in particular, other people).

In the first study, we tried to mimic an information retrieval situation in which subjects were searching for a particular item, knew a lot about it, but could not specify it by name. For example, one might know that the local computer has a program that accesses the Associated Press newswire, and further know that the program allows the user to look at just titles, just first sentences, stories containing specified keywords, etc., but have forgotten the program's name (e.g., ''window'') thereby making it impossible to access. In fact, estimates from work logs kept by engineers at Boeing showed that up to 70% of the time, people looked for specific things that they had seen before (e.g., standards, product manuals), but had forgotten (Lesk, 1980). The experimental task we studied was a communication game in which subjects were given a target word, and had to describe the target item in a way that would allow another person (or a hypothetical computer) to discover the word.

The items used in our study were selected from a pool of 50 familiar words chosen from 10 categories (see Table 1). These categories were not meant to be representative or exhaustive, but were selected merely to allow us to examine some possibly interesting aspects of natural descriptions—such as terseness and spontaneous use of negation. For example, one might expect that ''abstract words'' (e.g., number, science) would be harder to describe than concrete nouns (e.g., tiger, raisin), or that things with well-known ''opposites'' could easily (if not always accurately) be described using contrastives. For example, the target 'black,' can be very concisely described as: ''color, opposite of white.''

Most of the categories in Table 1 are self-explanatory, but the two ''not'' categories require some additional explanation. The words in these categories

TABLE 1
Categories and Examples of Items

1. CITIES—San Francisco; Seattle
2. PROPER NAMES—R. Jackson; A. Einstein
3. CLOTHING—sweater; mittens
4. ANIMALS—tiger; goat
5. FOOD—pear; raisin
6. HOUSEHOLD THINGS—toaster; TV
7. "ABSTRACT WORDS"—number; science
8. "OPPOSITES"—black; night
9. "NOT"—Empire State Building
10. "NOT"—Newsweek; motorcycle

were chosen in such a way that a simple description could reduce the set of likely alternatives to a fairly small set (including, of course, the target item); then a negation or other contrastive could be used to rule out all but the desired target item. For example, the target 'Newsweek' can be described as: "a popular weekly newsmagazine, *not* Time"; or the target 'Empire State Building' can be described as: "the tallest building in New York City, *other than* the World Trade Centers."

Three hundred thirty-seven New York University undergraduates were each given 10 words (one from each of the 10 categories shown in Table 1), and asked to get another person or a (fictitious) computer to discover the word. For half of the students, the first five items were to be described for computers and the last five descriptions for people. The opposite order was used for the other half of the subjects. The students were asked to be as clear and specific as possible in their descriptions. The only restriction was that they not use the target word itself in the descriptions. The descriptions were written on paper. The students also were asked to indicate whether they had experience with computers, and if so what the nature of that experience was. In the data summaries reported below, students with at least one computer course were classified as computer "experienced."

A sequel study was conducted to evaluate the effectiveness of the descriptions generated by the students. Twenty-five subjects (6 area homemakers and 19 Bell Telephone Laboratories employees) were each given 150 descriptions[1] randomly selected from those generated by the NYU students. They were asked to: a) guess the item being described, and b) indicate (on a 5-point scale) their confidence in each response.

Several characteristics of the descriptions generated in the main study were examined. We first looked at the number and types of words used. The mean num-

---

[1] Since there were only 50 items in our pool of items, this means that items could be described more than once. This could either: a) inflate the proportion correct (if people guessed an item on the basis of information in a previous definition), or b) decrease the proportion correct (if people tried not to repeat the same guess). In any event, this did not appear to have much of an effect, since the guessing accuracy was about the same for the first occurrence of descriptions of a particular target as for later descriptions (71% vs. 75%).

ber of words was 8.7, with a range from 5.2 (for 'bachelor') to 12.8 (for 'racquetball'). It is interesting to note that the descriptions of several common target words ('black,' 'necktie,' 'mitten,' and 'sweater') required many words (more than 11.0). Somewhat surprisingly, the category of "abstract words" had the second lowest group mean (8.0).

There was no significant main effect of whether the descriptions were intended for people (8.76 words) or computers (8.58 words) ($F(1,49) < 1.0$). There was a statistically significant effect of computer experience ($F(1,49) = 14.49, p < .001$); with computer-experienced people being more terse than computer novices (8.33 vs. 9.01 words). More interestingly, there was a reliable interaction between the level of computer experience of the describers and whether the descriptions were intended for people or computers ($F(1,49) = 6.54, p < .025$). This result is shown in Table 2.

TABLE 2
Mean Number of Words Used in Descriptions

	Intended Audience	
	To Person	To Computer
Person Providing Description		
No Computer Experience:	8.86	9.16
Some Computer Experience:	8.67	8.00

When communications were intended for computers, people with computer experience were relatively more terse, and non-experienced people were relatively more verbose than when communicating with people. Perhaps experienced computer users think computers need less explicit specification of default contextual conditions than do ordinary people; and, conversely, people without computer experience may believe that machines need more contextual information than people. However, there was no simple relationship between verbosity and effectiveness (i.e., the guessing accuracy as indicated by the sequel study), and the range in guessing accuracy for the four conditions shown in Table 2 was small (77.9% to 82.3%). People were somewhat more successful in guessing the target items when the descriptions had been provided by people without computer experience (81.2% vs. 78.5%), but this difference fails to reach statistical significance ($F(1,49) = 2.68, p \approx .11$).

In addition to these more quantitative aspects of students' descriptions, there were several interesting qualitative aspects of the data. A surprisingly large proportion of the descriptions were rather vague, in the sense that the class of objects referred to by the description (the denotative class) was fairly large. For example, a description like: "tall building" is not a very precise description of the target 'Empire State Building'. We identified all instances in which the description clearly applied equally well to many objects other than the target. About 20% of all the descriptions were vague by this standard. Another more objective measure of precision of the description was the proportion of correct guesses made by the subjects who read the descriptions. Interestingly, in spite of the fact that the class of objects referred to was fairly large, guessing accuracy (in the sequel study) was 80%. One interpretation of the general lack of specificity is that human describers presuppose that one member of a large class will be guessed in the absence of additional information or further specification, i.e., that the recipient will tend to return the one most representative or dominant item in the class of items specified by the description. For example, if the most representative or typical member of the class "popular weekly newsmagazine" is 'Newsweek,' there is no need (when communicating with other people) to specify additional information. Thus the surprisingly high guessing accuracy might simply reflect the typicality of our target items.

Subjects rarely used negations, despite the inclusion of several items (items in the two "not" categories, among others) that easily allowed such specification. Subjects used negations fewer than 25% of the time for items in the "not" categories. Descriptions (for the target item 'Newsweek') like "a popular weekly magazine, *not* Time" were rare. More common were definitions of the form "popular weekly magazine" (not specific); or "popular weekly magazine, published by the Washington Post" (superordinate plus characteristics, see below).

What then *do* subjects do when attempting to obtain information? By far the most common form of specification (occurring in approximately 60% of the descriptions) was to give a superordinate of the target item followed by a few distinguishing characteristics or features. For example, San Francisco was commonly described in terms like the following: "large city in California, noted for its trolley cars and Chinatown." Another common form of specification (used in about 15% of the descriptions) was a list of subordinates or other associates of the target item. In describing the target item motorcycle, subjects often said things like: "Harley Davidson, Suzuki, and Kawasaki are examples of this." While there are some interesting consistencies in the form of natural-item specifications, this is only a beginning. Much more work needs to be done to explore when each of these specification methods is likely to be used, when it can successfully be used, etc.

To summarize the results of this study, we've seen that there are a few common and preferred forms of natural-object specification. First, is specification in terms of a superordinate plus features. This may be a natural way to specify a set of features, i.e., by a conjunction of keywords, with items being given in order of priority or weight. Second, is specification by lists of associates or subordinates. Such lists might be the way that people naturally specify features that are common to all items listed. The recipient person or computer needs to be able to extract the common features and superordinate category name (e.g., motorcycle) from the set of exemplars.

While there are interesting possibilities for system design suggested by these descriptions, it is important to realize that we have no evidence as yet as to whether systems which allowed access using these more "natural" query forms would be easier to use. We do, however, know that without training, people do not use Boolean expressions much more complicated than simple conjunctions, nor are the expressions they do use very precise. Instead, people specify items vaguely (definitions often apply to several objects other than the intended target) in terms of superordinates and distinguishing characteristics, or by lists of associates.

3. Study 2

In another study of natural-command terminology, Landauer, Galotti, and Hartwell (1981) went one step further. They first identified spontaneously favored command words, and then *tested* the value of the resultant natural-command lexicon in the initial stages of learning to use a simple text-editing system. Text editing is a particularly interesting domain of study, because it is often the first exposure that computer novices have to computerized systems. Furthermore, there are frequent complaints from new users of such systems about the "unnatural" and technical jargon required.

In order to test the hypothesis that systems would be less difficult to introduce to novices if a more natural and familiar lexicon were used, Landauer et al. first developed a method for identifying natural-command words for a text-editing task. They obtained manuscripts that had been marked up by their authors for correction. From these, they developed a taxonomy of commonly requested editing changes. This taxonomy is summarized in Table 3.

TABLE 3
Taxonomy of Text-Editing Changes

$$\begin{bmatrix} \text{Insert} \\ \text{Delete} \\ \text{Replace} \\ \text{Move} \\ \text{Transpose} \end{bmatrix} \times \begin{bmatrix} \text{Blank} \\ \text{Character} \\ \text{Word} \\ \text{Line} \\ \text{Paragraph} \end{bmatrix}$$

There are five basic operations (insert, delete, replace, move, and transpose) which can be applied to each of the five units (blank, character, word, line, and paragraph), resulting in 25 types of changes.

Next, they recruited 48 people representative of potential new users of computer text-editing systems, secretarial students and high school students with typing experience, but no previous computer-related experience. These subjects were given a sample text with author's marks representing two occurrences each of the 25 possible types of changes. They were asked to prepare a typed list of instructions to someone else who was actually going to make the changes and who did not have the marks. (See also, Black & Sebrechts, in press.) Landauer et al. hypothesized that words one typist uses to instruct another typist would provide "natural" command names for text-editing operations.

The main verbs of these descriptions were used for further analyses and are summarized in Table 4.

There are several interesting characteristics of these descriptions. In general, these characteristics can be seen by examining the frequencies presented in Table 4, but have also been verified and extended using more sophisticated similarity scaling analyses (see Landauer et al., 1981 for additional details). First, the commands generated by typists in this experiment differ from those found in popular line-oriented editors (e.g., the UNIX™[2] text-editor 'ed'). In fact, in all but one of the 25 cells, the UNIX text-editor command was not the name most frequently given spontaneously by the human subjects. The second interesting aspect of these descriptions is that, for the most part, the subjects didn't agree with each other. On the average, the three most popular names for each operation account for only 33% of the total number of responses. Bearing in mind this general lack of intersubject agreement, the data suggest that typists prefer "add" for the insert operation, "omit" for the delete operation, and "change" for the replace operation. The final two characteristics of the command descriptions went beyond simple vocabulary to the scope of commands. The computer-naive typists used different words to describe operations on blanks and characters, whereas most text editors treat them identically. Apparently, typists do not consider a blank a character. Finally, typists tend to use the same words to refer to operations on words and lines. Line-based editors, at least, treat these two scopes differently. More specifically, the typists use the word "omit" for both within-line and whole-line operations, whereas in 'ed', "substitute" is used to remove a word, and "delete" to remove whole lines.

Having found a variety of discrepancies between the way potential users and current systems describe common editing operations, the next question is what to do about it. A tempting possibility is to change the command lexicon to agree with computer novices' preconceptions about the editing task with which they are familiar in manual form. Since the user-nominated words have well-known meanings, they should require minimal new learning. (This intuition seems to lie behind many recommendations for command-name choice, e.g., Norman, 1981.)

---

[2] UNIX is a trademark of Bell Telephone Laboratories, Incorporated.

TABLE 4
Frequency of User-Provided Names

Type of change	Blank		Character		Word		Line		Paragraph	
					Object					
Insert (put in text)	Substitute*	0	Substitute*	0	Substitute*	0	Append*	0	Append*	0
	Space	15	Change	20	Insert	25	Insert*	19	Insert*	8
	Put	9	Should be	17	Add	20	Add	24	Type	15
			Insert	11	Place	13	Type	17	Add	15
			Add	8	Put	13				
					Type	11			Put	11
	Other	72	Other	40	Other	14	Other	36	Other	47
	(22 types)		(17 types)		(9 types)		(13 types)		(11 types)	
Delete (remove text)	Substitute*	0	Substitute*	0	Substitute*	0	Delete*	6	Delete*	6
	Connect	13	Omit	15	Omit	31	Omit	34	Omit	32
			Spell	12	Take out	10				
	Change	11	Change	11	Delete	9	Take out	10		
							Eliminate	9		
	Other	72	Other	58	Other	46	Other	37	Other	58
	(31 types)		(20 types)		(15 types)		(9 types)		(14 types)	
Replace (new text in place of old)	Substitute*	0	Substitute*	1	Substitute*	4	Change*	11	Change*	0
	Add	13	Change	18	Change	22			Replace	6
	Insert	10			Replace	10				
	Put in	10	Replace	9						
	Change	9								
	Spell	9								
	Other	45	Other	68	Other	60	Other	85	Other	90
	(20 types)		(35 types)		(35 types)		(43 types)		(44 types)	
Move (change location of text)	Substitute*	0	Substitute*	0	Substitute*	0	Move*	2	Move*	8
	Should be	10	Change	16	Place	11	Put	12	Put	13
					Type	10				
			Put	13	Put	9	Type	9	Place	9
			Should be	17						
	Other	86	Other	50	Other	66	Other	73	Other	66
	(43 types)		(25 types)		(36 types)		(40 types)		(36 types)	
Transpose (interchange locations)	Substitute*	0	Substitute*	0	Substitute*	0	Move*	0	Move*	3
	Change	8	Change	17	Reverse	12				
			Should be	17						
			Switch	14	Switch	11			Switch	16
			Spell	10	Type	9	Type	5		
	Other	88	Other	38	Other	64	Other	91	Other	77
	(46 types)		(21 types)		(27 types)		(52 types)		(22 types)	

* = command name used in the UNIX text-editor 'ed'

However, there are at least two reasons why popular choices might *not* make optimal command names. First, popular terms tend to be referentially imprecise (cf. Study 1, above). Indeed, in the spontaneous instruction data, the same word was often used to refer to what are in 'ed' different operations. In the second column of Table 4, for example, "change" is the most frequently used term for all but one operation. Second, the command-name set should reflect the set of required computer operations. For example, if different syntax is required for operations on strings within lines and whole lines, different command names for the two cases might be advantageous. Insofar as the manual version of the text-editing task requires a different set of operations and syntax from those in the computer version, the set of command names generated in one environment might not be appropriate for the other.

The second part of the Landauer et al. study tested whether using user-nominated terminology would help people during the highly critical initial phases of learning a computerized text editor. (See also, Ledgard, Whiteside, Singer, & Seymour, 1980.) A variety of 'mini-editors' incorporating (and designed to investigate) the above differences were developed. Specifically, the editors used three operations (delete, append, and substitute) plus beginning and ending operations. There were 12 versions of the editor, which varied with respect to three parameters of command selection: a) the command vocabulary, b) command scope, and c) command length. A summary of the parameters of interest is shown in Table 5.

TABLE 5
Design of Text Editor Experiment

1. Vocabulary

Old	New	Random
delete	omit	allege
append	add	cipher
substitute	change	deliberate

2. Scope

Old:
Different words for whole-line and within-line operations.
New:
Same words used for whole-line and within-line operations.
EG:

Remove Command	*"old scope"*	*"new scope"*
whole line:	delete	delete
within line (word)	substitute/word//	delete/word/

3. Command Length

1. whole word
2. abbreviated to first letter

There were three sets of command vocabularies—the spontaneous user names (omit, add, change); the UNIX text-editor names (delete, append, substitute), and randomly chosen control verbs (allege, cipher, deliberate) which were unrelated to the editing operations, but matched to the existing 'ed' command names in frequency of use in the English language, and in length. The second variable, scope, had two levels. In the "old scope" condition, within-line changes and whole-line changes required different commands; in the "new scope" condition, the command was lexically the same for within- and whole-line changes. The final variable was command length which was either the entire command name or an abbreviation consisting of the first letter.

Sixty-five secretarial students and 56 high school students participated in the second phase of the study. Seventeen secretarial and eight high school students failed to learn enough to complete the test exercises used for data. The subjects (none of whom had previous computer experience) each spent about two hours studying a self-instructing manual and simultaneously doing a series of on-line learning and test exercises. Each of the 12 versions of the editor had a different version of the instruction manual, and eight subjects completed each of the 12 versions. The manuals for the 12 different conditions were nearly identical, with content differences limited to a few words and phrases which were systematically manipulated in the different conditions. The subjects went through six learning exercises in which new editing operations and their associated commands were introduced and practiced, followed by four test exercises which examined their mastery of these operations. A questionnaire about attitudes toward the computer-editing task was given at the end of the experiment.

The learning required in this experiment mimics fairly closely that of the first encounter of novices with real computer text-editing systems. It involves only a small subset of the "full" text-processing system, but brings the user to a point where actual work can be done. The learning was non-trivial, as evidenced by the fact that almost 20% of the subjects who started failed to finish within two hours. We have heard frequent reports that it is exactly these first few hours that are the most difficult hurdle for potential users who are not computer-oriented. Complaints about "strangeness," "unnaturalness," and "non-English" interaction are said to be most prevalent at the beginning of text-editor training. Thus, if the difficulty of learning text-editors is to be attributed to "unnatural" command names, there seems every reason to believe that command-name differences should have a large effect during the initial training period investigated in this experiment, and for just such an initial set of commands as it studied.

The mean total times to complete the four test exercises (as a function of experimental condition) were examined. Several interesting and somewhat surprising results of these analyses are summarized in Table 6.

First, the time to perform the *test* exercises was *not* significantly influenced by variations in command names. Subjects performed about as well when they were learning to "allege," "cipher," and "deliberate" as when they were learn-

TABLE 6
Mean Time to Perform Four Test
Exercises on Various Versions
of Minimal Test-Editors

Vocabulary: (108)	
Old:	1799 secs
New:	1801 secs
Random:	2039 secs
Length: (88)	
Long:	1816 secs
Short:	1936 secs
Scope: (88)	
Old:	1657 secs
New:	2094 secs

NOTE: Standard errors about the mean are
given in parentheses.

ing to "add," "change," and "omit" ($t(62) = 1.0$ for the largest pairwise comparison). Interestingly, there were very few complaints about the command names "allege," "cipher," and "deliberate." (In fact, the only person who complained vigorously was the experimenter who was often quite confused when asked questions of the sort: "What's wrong? I've alleged this twice and tried to cipher it, but nothing seems to work.") However, the post-session questionnaire revealed some subjective preferences for the more appropriate terms.

There is an important caveat about interpreting these results. These findings do not prove that meaningfulness and appropriateness will never have important effects on learning, only that they do not strongly influence initial learning and use of a very small set of commands. In fact, there is good reason to believe that appropriateness would be important with larger lexicons or when long-term recall was required. However, these results do show that natural words for command names, by themselves, do not seem to provide substantial benefit during the highly critical first few hours of introduction to a computerized text-editing environment.[3]

Surprisingly, the time differences between abbreviated and long command names were not statistically significant ($F(1,94) < 1$). However, an examination of the practice function showed that abbreviated command names were slightly more time-consuming to use at first (in spite of the fact that fewer keystrokes were required), but became significantly less so after time, (confirmed statistically by a significant interaction between command length and test exercise-number; $F(1,72) = 4.22, p < .05$).

---

[3] Ledgard et al., found superior initial learning of what they called an "English-based" over an "arbitrary" text-editing command set. However, the two systems differed primarily, and dramatically, in their required syntactical constructions, so there is no direct contradiction with our conclusions.

One possible explanation for this result is that typing out the entire name forces users to learn the *names* of the commands, which might, in turn, help them learn and remember how to perform the associated editing operation.

The scope manipulation was also effective. The old scope, in which different commands were used for within-line and whole-line operations, was significantly easier to use; 1657 sec versus 2029 sec for the four test exercises, respectively ($F(1,72) = 12.05$, $p < .001$). This result contrasts with subjects' preferences for using the same word to describe both types of operations. Our interpretation of this result depends on the fact that the text editor requires the use of different syntactic constructions for operations on whole lines and within lines (see Table 5). The old scope matches this system requirement; its command names also differentiate between whole- and within-line operations. Thus, different command words may help the user differentiate between the system's different syntactic requirements. Clearly, these results imply that if "naturalization" is going to help novice users over the initial hump, it will have to go further than lexical familiarity or surface semantics.

Note also that the large—almost 25%—and statistically highly significant effect of scope differences on learning time provide assurance that the experiment was sufficiently powerful to detect at least some effects. The contrastingly small effect of command-name choice is thus made harder to dismiss.

The results of the editor command studies can be summarized as follows. First, experienced typists who are computer novices do not use the same language as system designers do to describe text-editing operations. And, for the most part, they don't even agree with each other. Thus one person's obvious command name may not be another's (see also, Jackson & Tschergi 1981; Ledgard et al., 1980; and Furnas, Landauer, Gomez, & Dumais, 1982). Second, the selection of popular words from the non-computer environment is not a sufficient strategy for choosing command names, and probably reflects a far too simple and undifferentiated view of the processes and problems. Deeper issues, such as command specificity and matching the structure of differences in operations and syntax to the structure of command semantics, are probably involved, and are much in need of further study.

4. General Summary

We began by considering the hypothesis that computer systems would be more congenial and easier to learn if the communication protocol and lexicon were more 'natural' and familiar to naive users. In the first investigation, we saw that there were several interesting and fairly consistent characteristics of people's natural object specifications. Students rarely used negations, or other Boolean expressions more complicated than simple conjunctions. Instead, they attempted to obtain information using one of two general strategies. The most common form of specification (occurring about 60% of the time) was to give a superordinate of the target item followed by a few distinguishing characteristics. Another common form of specification (used about 15% of the time) was a list of subordinates or other associates of the target. While these suggest some interesting possibilities for computer query languages, they have yet to be implemented and tested in an actual information retrieval system.

The second series of experiments looked at how to choose command names. One hypothesis was that user-nominated command names have familiar, known meanings, and thus should require minimal new learning. This idea now appears to have been far too simple. While it is true (at least under some circumstances) that familiar items are easier to remember, the catch is that there is evidently much more to learning a text-editing system than remembering command names. Command names have to be applied to carefully specified situations. The popular names nominated by subjects may not have the necessary specificity. One possible compromise between the advantages of familiar words and the need for specificity might be to choose relatively low frequency words (that are specific) but also recognizable. Another problem is that of appropriately matching command syntax to the underlying operations of the system. Untrained users cannot be expected to generate command names conforming to these requirements, but system designers have the necessary information. Furthermore, the words designers use in trying to be precise rather than popular may also meet the criteria of low (but sufficient) familiarity and recognizable meaning. These two hypotheses may explain why the command names chosen by the designers of 'ed' were learned as rapidly as those generated spontaneously by the typists. While the problem of selecting ''natural'' protocols and vocabularies is a more complicated problem than it may have first appeared, experimental analyses will hopefully help sort out relevant variables.

5. REFERENCES

Ackroff, J.M. Unpublished paper, Bell Laboratories, August 1979.

Allen, R.B. Cognitive factors in human interaction with computers. In A. Badre and B. Shneiderman (Eds.) *Directions in human/computer interaction*. Ablex Publishing Corp., 1982.

Black, J.B. & Sebrechts, M. M. Facilitating human-computer communication. *Journal of Applied Psycholinguistics*, in press.

Braine, M.D.S. On the relation between natural logic of reasoning and standard logic. *Psychological Review*, 1978, *85*, 1–21.

Furnas, G.W. *Psychological structure in information organization and retrieval: Arguments for more considered approaches, and work in progress*. Paper presented at the Human Computer Interface Symposium, Georgia Institute of Technology, March 1981.

Furnas, G.W., Landauer, T.K., Gomez, L.M. & Dumais, S.T. *Statistical semantics: Analysis of the potential performance of keyword information access systems*. Unpublished paper, Bell Laboratories, 1982.

Jackson, M.D. & Tschergi, J. *The nature of user-generated commands for interacting with a computer.* Paper presented at the Human Computer Interface Symposium, Georgia Institute of Technology, March 1981.

Landauer, T.K., Galotti, K.M. & Hartwell, S. *A computer command by any other name: A study of text editing terms.* Unpublished paper, Bell Laboratories, 1981.

Ledgard, H., Whiteside, J.A., Singer, A. & Seymour, W. The natural language of interactive systems. *CACM, 1980, 23* (10), 556–563.

Lesk, M.E. Another view. *Datamation,* November 1981, *27* (12), 146.

Lesk, M.E. Personal communication, 1980.

Norman, D.A. The trouble with UNIX. *Datamation,* November 1981, *27* (12), 140–150.

Reisner, P. Human factors of data-base query languages: A survey and assessment. *Computing Surveys,* 1981, *1,* 13–31.

Shneiderman, B. A note on human factors issues of natural language interaction with data-base systems. *Information Systems,* 1981, *6,* 125–129.

Stevens, P. & Shneiderman, B. Exploratory research on training aids for naive users of interactive systems. *Proceedings ASIS Conference,* October 1981, Washington, DC.

Thomas, J.C. Psychological issues in data base management. *Proceedings of the Third International Conference on Very Large Data Bases.* New York: IEEE, 1977, 169–184.

Thomas, J.C., & Gould, J.D. A psychological study of Query-by-Example. *Proceedings of the National Computer Conference. 1975, 20,* 237–261.

6

Abbreviations for Automated Systems: Teaching Operators The Rules*

S. L. EHRENREICH[1]
THEODORA PORCU

U.S. Army Research Institute for
the Behavioral and Social Sciences
5001 Eisenhower Avenue
Alexandria, VA 22333

One way to improve performance on abbreviations is to inform operators of the rules used to generate the abbreviations. This was tested in a series of rating, encoding, and decoding experiments which compared the benefits of truncation versus contraction and fixed versus variable length abbreviations. In addition, the advisability of representing common suffixes (ING, ED, S) in abbreviations was tested along with a technique for dealing with the problem of a simple rule generating the same abbreviation for more than one word. Based upon the results of these experiments, guidelines for generating abbreviations are presented.

1. INTRODUCTION

Abbreviations are universally employed in computers to expedite data entry and reduce the space (field) occupied by a message. But all abbreviation techniques are not equally effective, and a number of factors must be weighed in selecting among them. A prime consideration is whether the main purpose for having abbreviations is to facilitate data entry or to reduce message length. In the former case, the user/operator (these terms are used interchangeably in this paper) needs

---

* The views expressed in this paper are those of the authors and do not necessarily reflect the views of the U.S. Army or of the U.S. department of Defense.
[1] Now at Bell Laboratories, Whippany Road, Whippany, NJ 07981

only to *encode* words (i.e., convert them into their abbreviations) prior to typing them. But, if abbreviations are used to reduce message space, then operators must *decode* the abbreviations (i.e., translate them into words) in order to understand the message. Abbreviations that are easy to decode may not be easy to encode.

Another factor which may influence the choice of an abbreviation technique is the number of words which are to be abbreviated. For example, command languages (languages used to execute a program conveniently, e.g., job-control languages, text editors) typically have small vocabularies. On the other hand, query languages (Ehrenreich, 1981) and fill-in-the-blank (i.e., form-filling) dialogues may have over 100 words in their vocabularies. An abbreviation technique that is suitable for a small vocabulary may not be suitable when the vocabulary is large. The technique of "minimum-to-distinguish" (also called "command completion" and "autocompletion") is an example.

Under the system of minimum-to-distinguish, an operator enters only the first few letters of a word. But enough letters must be entered to make the abbreviation unique. For vocabularies containing 10 words, the first one or two letters of a word will usually suffice. But if a vocabulary contains 100 words, then the number of initial letters that are needed will vary greatly from one word to another. Thus, the minimum-to-distinguish technique requires that operators learn how many letters are needed to abbreviate each word.

Still other factors which must be considered in choosing an abbreviation technique are operator training, experience, and frequency of interaction (i.e., how often the operator uses the system and its abbreviations). Most any reasonable abbreviation technique will suffice for experienced operators who are constantly working with the system. But when operator turnover is frequent (as it is true for military systems) or if the system is designed to support non-dedicated users, the shortest abbreviations might not be the best.

Finally, *speed*-accuracy and *space*-accuracy trade-offs should be considered. For instance, the longer an abbreviation is (and thus the more space it occupies), the greater the likelihood that it will be interpreted correctly. Likewise, the more letters that an operator enters when using the minimum-to-distinguish technique, the greater the likelihood that the abbreviation will be unique and therefore acceptable. However, this increase in accuracy occurs at a cost in the number of key-strokes and the amount of space occupied on the display.

2. BACKGROUND

There exist numerous techniques for generating user-oriented abbreviations. These include techniques for producing abbreviations that are phonetic approximations of the original words (Schneider, Hirsh-Pasek, & Nudelman, 1981) and techniques for producing "natural" abbreviations (Ackroff & Streeter, 1981;

Streeter, Ackroff, & Taylor, 1980). The latter refers to abbreviations which are similar to the ones people create for themselves. Still another abbreviation technique can be found in McBride, Lambert, and Lane (1981). Their abbreviation algorithm selectively deletes the less "important" letters in a word.

The two most common methods for creating abbreviations are truncation and contraction (Hodge & Pennington, 1973). The abbreviation technique of *truncation* involves retaining the first few letters of a word and deleting the remaider. Moreover, the number of letters that are retained may either be constant (*fixed-length* abbreviations), or it can vary depending upon the length of the original word (variable-length abbreviations).

In *contraction* abbreviations, some specified set of letters (usually vowels) are deleted starting at the right end of a word and progressing to the left. However, the first letter of the word is never deleted. (For Hodge & Pennington, 1973, the last letter of the word is also always retained.) Again, these abbreviations can be either fixed or variable in length. The reason for deleting letters in right-to-left order is that for some words, deleting all of the vowels might result in too short an abbreviation. An example is the word CREASE, whose contraction (vowels removed), fixed-length (four letters) abbreviation is CRES. Examples of abbreviations formed by the truncation and contraction techniques are shown in Table 1.

A set of experiments by Moses, Mendez, and Ehrenreich (1980) compared performance on abbreviations generated by some of these techniques. In these experiments, participants studied 90 word-abbreviation pairs (e.g., CAPTAIN-CPT), seeing each pair either one, three, or six times (repeated items were *not* blocked within the list). Three techniques were used to generate the abbreviations: truncation-variable length, contraction-variable length (vowels and H, W, Y were removed), and abbreviations proposed for Army use (these abbreviations followed no systematic pattern). For each participant, one third of the abbreviations were formed by each technique, and the abbreviations from the three techniques were randomly placed within the stimulus lists. Thus, participants could not discern an abbreviation rule when performing the task. The stimulus pairs were individually projected onto a screen and each time that they appeared, participants copied them using pencil and paper.

Following the study phase, participants were divided into groups. One group was tested on its ability to encode all 90 words (produce the abbreviation when given the word), while the other group was tested on its ability to decode all 90 abbreviations (recall the word when given the abbreviation).

For encoding, there was no significant difference between Army abbreviations and truncation abbreviations, and both were significantly superior to contractions abbreviations. Performance on the Army and truncation abbreviations ranged from 31% correct for stimuli seen only once during the study phase to 62% correct for stimuli seen six times. Decoding performance, however, showed no significant differences between the three abbreviation techniques. Abbreviations which had been seen only once during the study phase were correctly decoded 70% of the time, and performance increased to 88% correct for stimuli that had been seen six times.

Poor encoding performance, as in the Moses et al. (1980) experiments, can be attributed to the absence of a systematic relationship between words and their abbreviations. This forces participants to rely upon rote memory in order to perform the task. That is, they have to learn every abbreviation as well as its association to a word. But if the abbreviations were generated by a simple rule which the participants understood, then the memory load would be greatly reduced. Even for decoding, knowledge of the abbreviation rule might help the participants in recalling the correct word.

If knowing a rule helps participants to process abbreviations, then the simpler the rule, the greater the benefit. But there lies a problem. The simpler the rule, the greater the likelihood that it will produce the same abbreviation for more than one word. For example, when using the truncation-fixed length rule to abbreviate TRANSLATE and TRANSPORT, the resulting abbreviation, TRAN, is identical for both items. To circumvent this problem, some abbreviations will have to deviate from the rule (i.e., deviant abbreviations).

When deviant abbreviations occur, rote memorization is reintroduced into the task. Even if a simple secondary rule exists for generating the deviant abbreviations, users have to learn which words are abbreviated using the secondary rule. By default, all of the other abbreviations are formed by the primary rule. Although the operator is still required to do some learning, the amount of learning is small.

In an effort to determine the value of teaching operators abbreviation rules, three experiments were performed (a pilot study is reported in Moses et al., 1980). These experiments tested different rules to determine their ease of use. In addition, the experiments examined the problem of deviant abbreviations. Finally, a method for representing common word suffixes (i.e., ING, ED, and S) was created and tested. The experiments did not, however, investigate acronyms, e.g., "radar," "snafu," or the practice of stringing together the initial letters of different words, e.g., "USA," "IBM."

3. EXPERIMENTAL APPROACH

Three experiments were performed to investigate: (1) people's subjective ratings of abbreviations; (2) their ability to encode words; and (3) their ability to decode abbreviations.

Each experiment considered four variables: (1) abbreviation technique (truncation or contraction); (2) abbreviation length (fixed or variable); (3) the effect of incorporating endings (ING, ED, and S) into abbreviations; and (4) the effect of intermixing abbreviations which were generated by an abbreviation rule with those which were not (deviants). The same design was used for each experiment and it is shown in Figure 1.

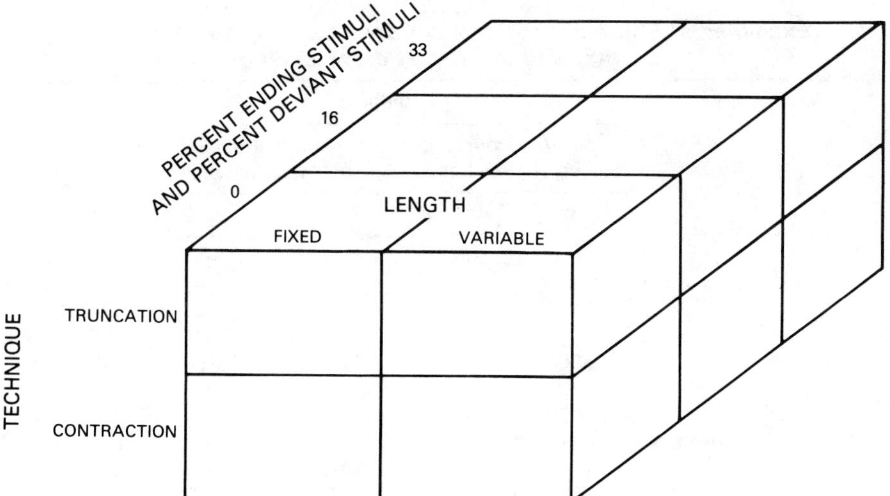

FIGURE 1. Design used for the rating, encoding, and decoding experiments.

Endings were incorporated into abbreviations by appending either a G, D, or S, to those words having ING, ED, or S as a suffix. For abbreviations formed by the truncation technique, the first few letters of the word were retained as previously described, but then the appropriate suffix letter was added. A similar statement holds true for abbreviations formed by the contraction technique. This rule made abbreviations of words with the appropriate suffixes one letter longer than abbreviations of equal length words without the suffixes (see Table 1). Either 0, 16, or 33% of the abbreviations in each condition incorporated endings.

Deviant abbreviations were ones that did not obey any of the four rules, i.e., truncation-fixed length (T-F), truncation-variable length (T-V), contraction-fixed length (C-F), or contraction-variable length (C-V). Instead, these deviant abbreviations were unsystematic though "reasonable." Deviant abbreviations never incorporated endings, and abbreviations that incorporated endings were never deviant. Whatever percentage of abbreviations in a list incorporated endings, then an equal percentage of *other* abbreviations were deviant (e.g., if 33% of the stimuli incorporated endings, than *another* 33% were deviant). Although the "endings" variable and the "deviation" variable always co-occurred, they are quite distinctive. Thus, significant effects found in the data could always be reasonably attributed to either one or the other, as will be seen.

Each participant performed in only one of the 12 conditions and remained in that condition throughout the three experiments (rating, encoding, and decoding). Before performing the encoding and decoding tasks, but not the rating task, participants were informed of the manner by which the abbreviations were formed.

TABLE 1
Examples of Abbreviations Which Vary According to Technique,
Length, and Incorporation of Endings

		Endings			
		Not Incorporated Abbreviation Length		*Incorporated*[1] Abbreviation Length	
		Fixed	*Variable*[2]	*Fixed*	*Variable*[2]	*Original Term*
	Truncation	MILI	MILI	-----[3]	-----	MILITARY
		VEGE	VEGET	-----	-----	VEGETATION
		ARTI	ARTIL	-----	-----	ARTILLERY
		ATTA	ATTAC	ATTAG	ATTACG	ATTACKING
Abbreviation		FORW	FORW	FORWS	FORWS	FORWARDS
Technique						
	Contraction	MLTR	MLTR	-----	-----	MILITARY
		VGTT	VGTTN	-----	-----	VEGETATION
		ARTL	ARTLL	-----	-----	ARTILLERY
		ATTC	ATTCK	ATTCG	ATTCKG	ATTACKING
		FRWR	FRWR	FRWRS	FRWRS	FORWARDS

[1]Endings are incorporated into abbreviations by appending a G, D, or S to the abbreviations of words having ING, ED, or S as their suffix.
[2]The variable length rule is:

if length of word is:	*then length of abbreviation is*:
5 letters or less	do not abbreviate
6-8 letters	4 letters
9 letters	5 letters

In computing the length of a word, all letters are counted, including suffixes.
[3]Where dashes (-----) appear, the abbreviation is the same as when endings are not incorporated.

4. RATING EXPERIMENT

In this experiment, participants were shown word-abbreviation pairs and asked to rate the "goodness" of the abbreviations. The abbreviations were formed by either the T-F, T-V, C-F, or C-V method. In addition, some of the abbreviations incorporated endings and some were deviant.

4.1 Method

Participants. In response to a request for participants, 144 military, enlisted personnel were assigned to the experiment. They came from varied backgrounds and occupational specialties. Immediately prior to this experiment, the participants had performed in a similar set of experiments where they rated, encoded, and decoded stimuli with*out* knowing the rules used to generate the abbreviations.

Materials. A set of terms (e.g., PENETRATE, NUCLEAR HAZARD) used on military command and control systems served as a source of stimuli for this experiment. In order to meet certain experimental requirements, some additional terms were added to this pool. All terms consisted of either one or two words, each word being five letters or longer in length.

From the pool of terms, 16 lists were constructed; the purpose for having more lists than conditions is explained below. Each list consisted of 72 term-abbreviation pairs (e.g., PENETRATE—PENE). When a term consisted of two words, each word was abbreviated individually. The abbreviation for the term was then formed by joining the abbreviations for the two words with a space between them (e.g., CIVILIAN BRIDGE—CIVI BRID).

Four lists were created by using *one* of the abbreviation methods (T-F, T-V, C-F, or C-V) to form all of the abbreviations in a list. These lists were labeled ''(0)'' since they contained no deviant abbreviations and no abbreviations which incorporated endings (although some of the words did end in ING, ED and S).

Another four lists, labeled ''(16A),'' were created out of the four (0) lists. This was done by having endings incorporated into 16% of the abbreviations, and by making another 16% of the abbreviations deviant. To assure the representativeness of the 16% condition, a second set of four lists, labeled ''(16B),'' was created. These lists were identical to the (16A) lists, except that the terms having endings and the terms with deviant abbreviations were different from those in the (16A) lists.

Finally, out of the four (0) lists, four lists labeled ''(33)'' were constructed. In these lists, 33% of the abbreviations were deviant and another 33% incorporated endings. In both the 16 and 33% conditions, an asterisk always appeared alongside each word-abbreviation pair containing a deviant abbreviation (e.g., *DISPLACE-DISL). Examples of the four lists (i.e., 0, 16A, 16B, and 33) formed by using the T-F abbreviation method are shown in Table 2. Lists formed by the T-V, C-F, and C-V methods were similarly constructed.

The 72 term-abbreviation pairs were typed on six pages. Alongside each pair was a rating scale ranging from 1 to 6. The digit 1 was labeled ''poor'' and the digit 6 was labeled ''excellent.''

TABLE 2
Sample Sections of Lists T-F(0), T-F(16A), T-F(16B), and T-F(33)

List T-F(0)	List T-F(16A)	List T-F(16B)	List T-F(33)
ARTILLERY-ARTI	ARTILLERY-ARTI	ARTILLERY-ARTI	ARTILLERY-ARTI
CIRCLE-CIRC	CIRCLE-CIRC	CIRCLE-CIRC	*BATTLE-BATL
REPEATED-REPE	POSITIONED-POSID	SPRAYED-SPRAD	SPRAYED-SPRAD
MILITARY	MILITARY	MILITARY	MILITARY
PEOPLE-MILI PEOP	PEOPLE-MILI PEOP	PEOPLE-MILI PEOP	PEOPLE-MILI PEOP
SCREEN-SCRE	*BATTLE-BATL	*ENVISION-ENVN	*ENVISION-ENVN
DETONATE-DETO	DETONATE-DETO	DETONATE-DETO	POSITIONED-POSID

Procedure. Participants were randomly assigned to one of 16 groups, with the restriction that 12 participants were assigned to each of the four (0) and four (33) lists, while six participants were assigned to each of the four (16A) and four (16B) groups.

Participants were instructed to circle a number on the rating scale to indicate how well each abbreviation represented its corresponding term. In addition, they were told to remember the term-abbreviation pairs, for they would be tested on them later. Participants were not informed of the rules used to generate the abbreviations in their lists and were given about 20 minutes to perform the task.

4.2 Results and Discussion

Within each list, there were three categories of stimuli: (1) terms where the abbreviations followed a simple rule ("simple" stimuli), i.e., T-F, T-V, C-F, or C-V; (2) terms having ING, ED, or S as a suffix ("ending" stimuli); and (3) terms whose abbreviations deviated from the simple rules ("deviant" stimuli). Note that lists (0), (16A), (16B), and (33) each had ending stimuli although only the latter three lists incorporated the endings into the abbreviations. This is demonstrated in the third row of stimuli in Table 2.

Since lists (16A) and (16B) were similarly designed, the data from these two lists were pooled prior to the analyses for the three experiments. This is justified by the fact that eight of nine independent *t*-tests comparing performance of the two lists showed no significant differences.

Figure 2 shows mean ratings broken down by stimulus categories: simple, ending, and deviant. Each category has three factors: abbreviation technique, abbreviation length, and percentage of abbreviations which incorporated endings and which deviated from the simple abbreviation rule (the same percentage for both). A three-factor analysis of variance was performed on each stimulus category.

Rule-Generated Abbreviations. For abbreviations formed by a simple rule (simple stimuli), those formed by the contraction technique were rated slightly but significantly higher than those formed by the truncation technique (3.93 vs 3.56), $F(1,132)=5.57, p< .02$. However, ratings for simple stimuli were not affected by either the length of the abbreviations or the percentage of deviant abbreviations (or abbreviations with endings).[2]

Abbreviations Incorporating Endings. The mean rating for abbreviations that incorporated endings (endings stimuli) was higher with the contraction

---

[2] *F* ratios are not reported for main effects that were not statistically significant (i.e., $p> .05$). In addition, interactions are mentioned ony if they were statistically significant. Complete ANOVA tables are reported in Ehrenreich and Porcu (1982).

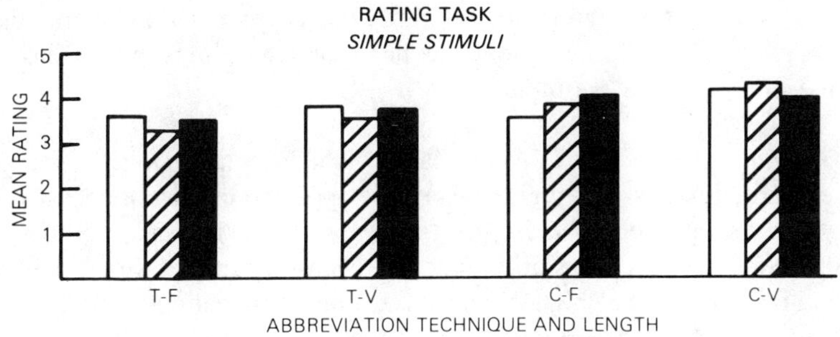

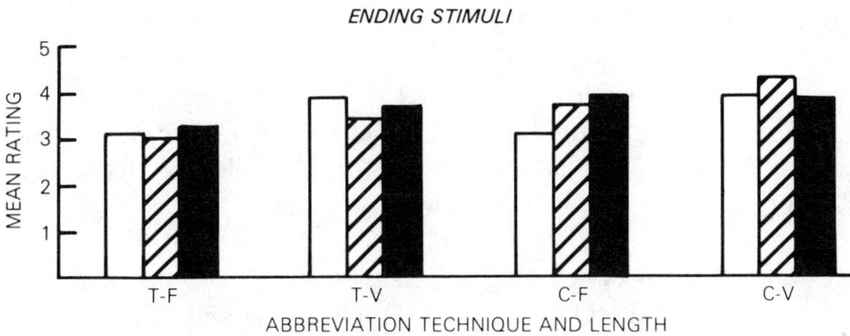

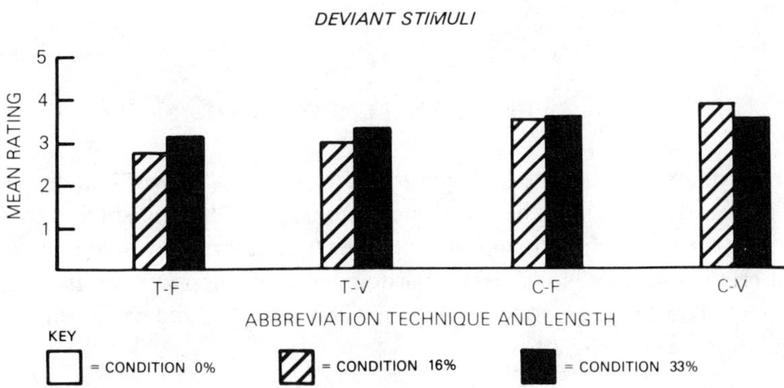

FIGURE 2. Mean rating of abbreviations by abbreviation technique, abbreviation length, and list condition.

method than with the truncation method (3.92 vs 3.34), $F(1,88)=8.59$ $p< .01$. However, there was no significant difference due to abbreviation length, and participants rated equally abbreviations that incorporated endings (conditions 16 and 33), and those that did not (condition 0).

Deviant Abbreviations. Abbreviations which deviated from the simple rule (deviant stimuli) were rated higher when they appeared in a list of contraction abbreviations as opposed to a list of truncation abbreviations (3.60 vs 3.04), $F(1,88)=8.09$, $p< .01$. This difference is probably an artifact. Since contraction abbreviations were rated higher than truncation abbreviations, this could have produced a bias towards higher ratings in the former type of lists. Abbreviation length and proportion of deviant abbreviations had no significant effect upon the ratings given deviant stimuli.

Rule-Generated Versus Deviant Abbreviations. For each of the eight lists in conditions (16) and (33), deviant abbreviations were rated lower than rule-generated abbreviations. Using a binomial distribution (i.e., sign test for matched pairs), the probability of this occurring by chance is $p = .008$, two-tailed test. Thus, within a list of abbreviations generated by a simple rule, abbreviations that deviate from that rule are not judged to be as "good."

Summary. Subjective ratings of abbreviations found contraction abbreviations to be slightly preferred over truncation abbreviations. On the other hand, there was no preference as to abbreviation length. Also, participants had no preference regarding whether or not an ending was incorporated into an abbreviation. As for abbreviations formed by a rule, they were rated higher than abbreviations generated in an unsystematic manner.

5. ENCODING EXPERIMENT

This experiment investigated people's ability to encode words after they had learned the rules used to generate abbreviations. After having studied the word-abbreviation pairs during the rating experiment, participants were taught the relevant abbreviation rules. In addition, participants in the appropriate list conditions were taught how to incorporate endings into abbreviations and were told that some abbreviations were deviant.

5.1 Method

Participants. The participants in this experiment were the same as those in the rating experiment. Each participant served in the same condition, e.g., T-F(0), C-V (16A), as he or she had served previously.

Materials. From each of the lists used in the rating experiment, half of the terms (but none of the abbreviations) were selected. Included in these terms were half of those having abbreviations which incorporated endings and half of those having abbreviations that were deviant. Thus 16 new lists, each containing 36 terms, were created. Terms whose abbreviations were deviant had an asterisk before them (i.e., *DISPLACE), as had been the case in the rating experiment.

The 36 terms in a list were typed on four pages. Alongside each term was a space in which participants could write the corresponding abbreviation.

Procedure. Participants were given written instructions on how to generate abbreviations (e.g., ''Generate abbreviations by retaining the first four letters of a word . . . .''). The instructions were unique to the participant's condition (T-F, T-V, C-F, or C-V). For participants in conditions (16) and (33), the instructions also described how endings were to be incorporated into abbreviations. In addition, these instructions mentioned that some of the terms had abbreviations which deviated from the simple rule and that each of these terms was marked with an asterisk. The participant was told that the correct abbreviations for these marked (i.e.,deviant) terms were the ones seen previously in the rating experiment. To help them understand the instructions, participants were given practice terms to encode. After completing the practice, they were tested on their ability to encode 36 items. During the test, participants were allowed to refer to the written instructions describing how to generate abbreviations. Participants were given approximately 20 minutes to read the abbreviation rules, perform the practice trials, and complete the test.

5.2 Results and Discussion

The encoding task was analyzed in the same manner as the rating experiment[3]. Figure 3 shows the mean percentages of correctly encoded terms.

Rule-Generated Abbreviations. For terms to be encoded using rules (simple stimuli), encoding performance was significantly better with the truncation technique than with the contraction technique (.87 vs .64), $F(1,132)=26.56, p<$ .001. This can be attributed to the fact that truncation is a simpler rule than contraction. In contraction, the participants must distinguish vowels from consonants, while in truncation, the participants need simply count off the first few letters. There were no significant differences due to abbreviation length or proportion of terms with deviant abbreviations. Although none of the two-way in-

---

[3] The encoding and the decoding experiments were analyzed using both raw scores and scores transformed by the arc sine transformation. For two interactions, an effect that was barely significant using the raw scores was not significant using the arc sine transformation. These two interactions are thus not reported. The statistics reported in the paper are based on the raw scores.

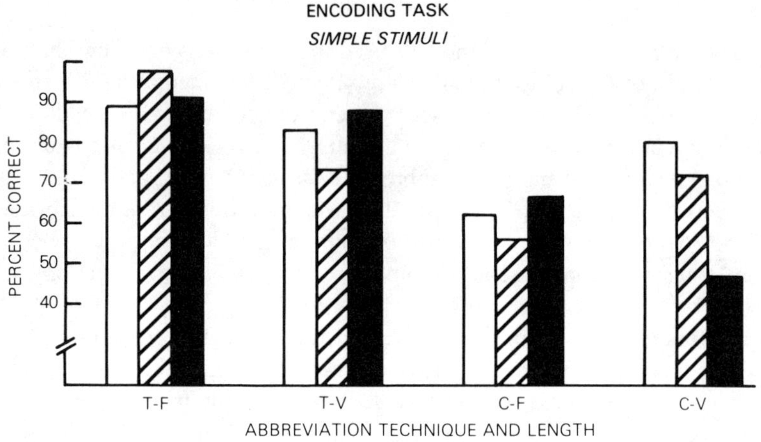

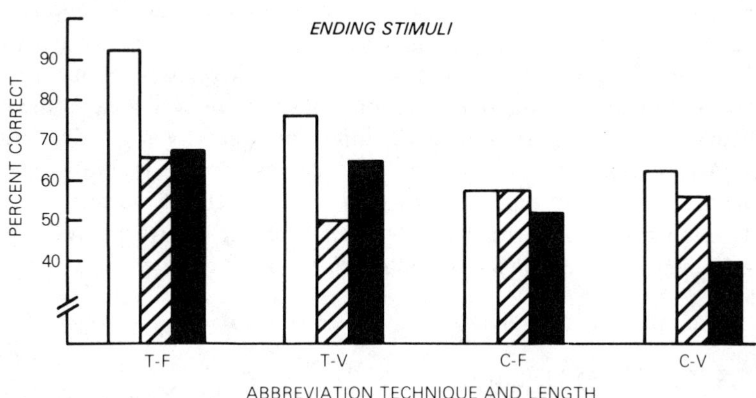

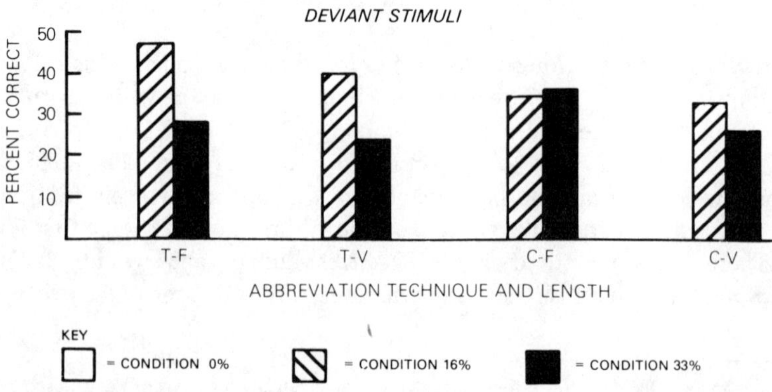

FIGURE 3. Mean encoding performance by abbreviation technique, abbreviation length, and list condition.

teractions were significant, there was a significant three-way interaction, $F(2,132)=3.69$, $p<.05$. The interpretation of such an interaction is unclear.

Abbreviations Incorporating Endings. When endings were incorporated into abbreviations, encoding performance was marginally better with truncation than with contraction (.62 vs .52), $F(1,88)=3.63$, $p<.10$. However, it was easier to ignore a word's ending (.72 correct in condition 0) than it was to incorporate it into an abbreviation (.57 correct in conditions 16 and 33), $F(2,132)=4.62$, $p<.02$. Finally, when encoding terms with endings, there was no significant difference due to abbreviation length.

Deviant Abbreviations. For terms whose abbreviations deviated from the rules (deviant stimuli), performance did not depend upon the proportion of deviant abbreviations in the list. Nor did it depend upon the length or technique used to abbreviate the other terms in the list. There was, however, a significant three-way interaction, $F(1,88)=7.05$, $p<.01$.

Rule-Generated Versus Deviant Abbreviations. For all of the eight lists in conditions (16) and (33), participants were better at encoding terms whose abbreviations followed a rule than in producing abbreviations that followed no rule. (This difference can be reasonably attributed to knowledge of the rule and not to orthographic or other differences between the two sets of abbreviations. See General Discussion.) Using the binomial distribution, the probability of this occurring by chance is $p = .008$, two-tailed test.

Summary. When participants knew the abbreviation rules, truncation abbreviations were easier to produce than contraction abbreviations. In addition, it did not matter if the abbreviations were fixed or variable in length. But performance was poorer when endings had to be incorporated into abbreviations. As for terms whose abbreviations were deviant, their presence did not affect the encoding of terms which followed a rule. However, terms having deviant abbreviations were marked with an asterisk. This will be discussed again later. Finally, the encoding of terms whose abbreviations followed a rule was superior to the encoding of terms whose abbreviations did not.

6. DECODING EXPERIMENT

The final experiment investigated people's ability to decode abbreviations when they knew the rules used to generate the abbreviations. Also examined was the performance on both deviant abbreviations and abbreviations which incorporated endings.

6.1 Method

Participants. The same participants performed in the decoding experiment as had performed in the prior two experiments. Each participant served in the same condition as he or she had served previously.

Materials. Each list used in the rating experiment had previously been split in half (see Encoding Experiment). The present experiment utilized the half lists not used for the encoding experiment. Only the abbreviations (and not the terms) from these half lists were used.

For each of the 16 conditions, a list of 36 abbreviations was formed. Each list contained the appropriate percentage of abbreviations which incorporated endings and the appropriate percentage of deviant abbreviations. Deviant abbreviations always appeared with an asterisk before them (e.g., *DISL). The items used in the decoding experiment were different from the items used in the encoding experiment. All items, though, had appeared in the rating experiment.

The 36 abbreviations for each list were typed on three pages. Alongside each abbreviation was a space for writing the term represented by the abbreviation.

Procedure. Participants were given approximately 15 minutes to decode the words. During the experiment, participants retained the instruction sheets given to them in the encoding experiment which described the rules for generating abbreviations.

6.2 Results and Discussion

Responses were scored in two ways. One method—strict scoring—considered a response correct only if it was identical to the word seen in the rating experiment. The second method—liberal scoring—considered a response correct even if it was misspelled or if it contained a different ending (e.g., ING, S, ED). As an example, consider the abbreviation PENE which had been paired with PENETRATE in the T-F condition during the rating task. If the participant's response to this stimulus was PENATRATES, the response was marked incorrect under strict scoring and correct under liberal scoring.

The decoding data were analyzed in the same manner as the data in the previous two experiments. Figures 4 and 5 show the mean percentage of abbreviations that were correctly decoded using liberal and strict scoring, respectively.

Rule-Generated Abbreviations (Liberal Scoring). Using liberal scoring, truncation and contraction abbreviations were decoded equally well (.63 and .59, respectively). Additionally, the ability to correctly decode a rule-generated abbre-

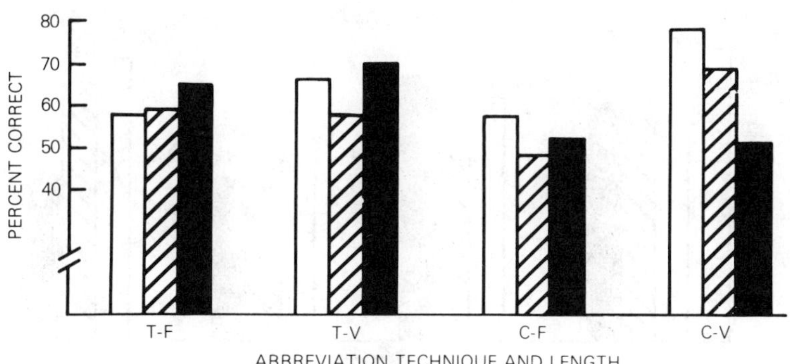

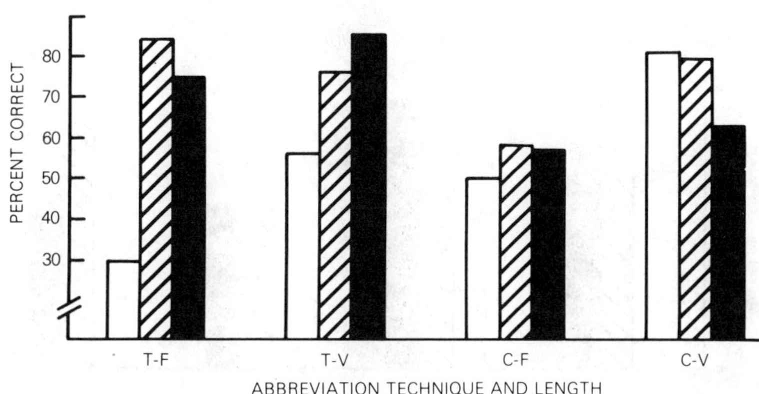

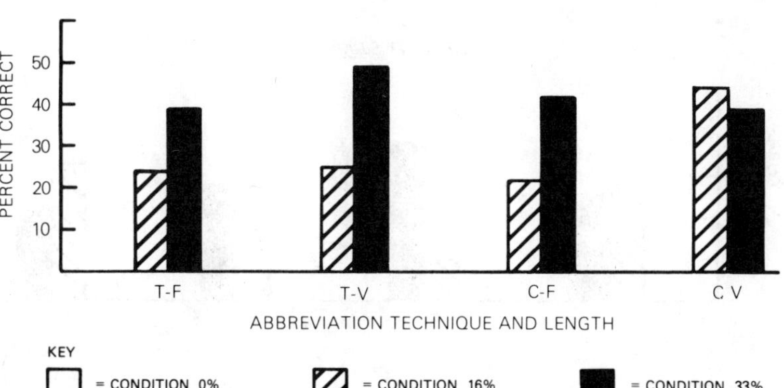

FIGURE 4. Mean decoding performance (liberal scoring) by abbreviation technique, abbreviation length, and list condition.

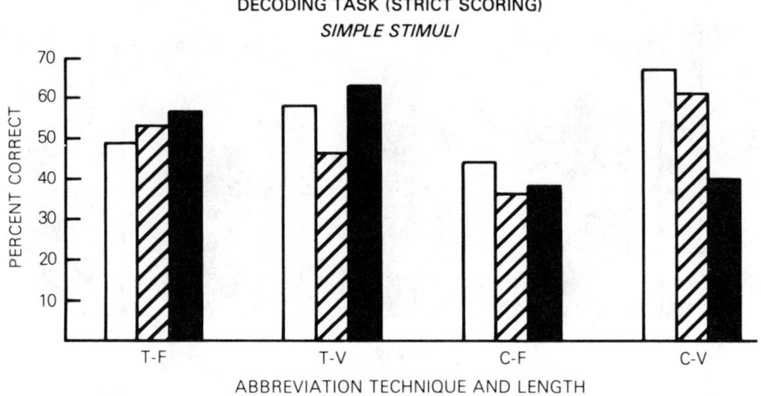

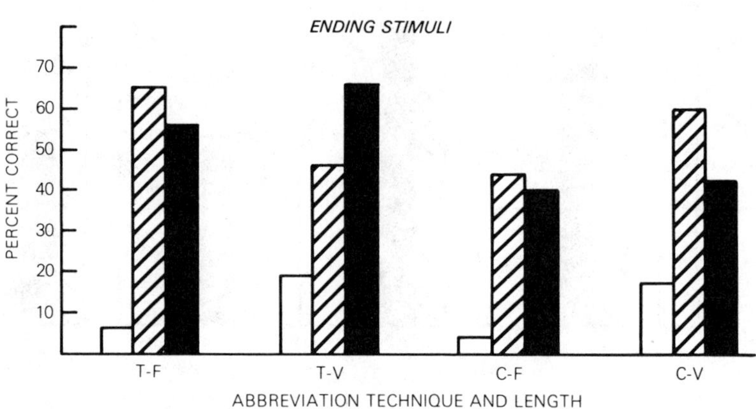

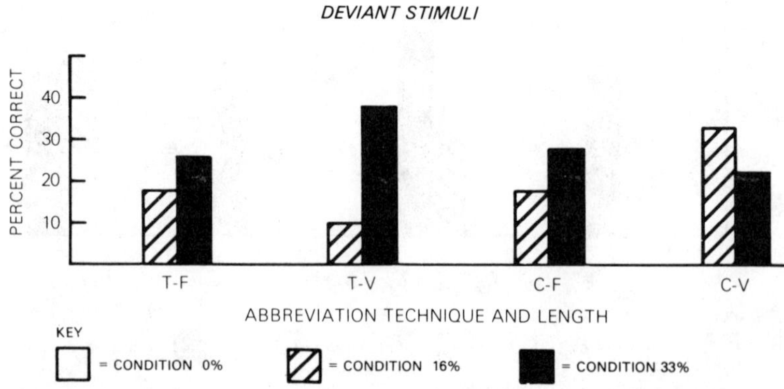

FIGURE 5. Mean decoding performance (strict scoring) by abbreviation technique, abbreviation length, and list condition.

viation did not depend on how many deviant abbreviations appeared in the list. However, variable-length abbreviations were slightly easier to decode than fixed-length ones (.65 vs .57, respectively), $F(1,132)=8.21$, $p<$ .01. This can be accounted for by the fact that variable-length abbreviations had as many or more letters than corresponding fixed-length abbreviations.

There was also a significant interaction between the rule used to generate abbreviations and the presence of deviant abbreviations, $F(2,132)=4.77$, $p<$ .01. Performance on contraction abbreviations was superior to performance on truncation abbreviations when there were no deviant abbreviations in the list (.68 vs .62 correct for lists 0). But the situation reversed for lists containing deviant abbreviations (.52 vs .68 correct for lists 33). Thus, truncation appears to be a better abbreviation technique when deviant abbreviations are also present.

Rule-Generated Abbreviations (Strict Scoring). When performance on rule-generated abbreviations was analyzed using strict scoring, all of the significant effects described above were again found. However, with strict scoring, there was a statistically significant effect due to abbreviation technique. Overall, abbreviations formed by truncation were easier to decode than abbreviations formed by contraction (.54 vs .48), $F(1,132)=4.53$, $p<$ .05. This slight superiority of truncation appears to result from the fact that it is easier to correctly spell a word when given its first few letters as opposed to its first few consonants. When spelling, people may be more likely to err on a vowel than on a consonant, as the former are less distinctive. Since truncation abbreviations contain vowels, words decoded from them would more often be spelled correctly.

To probe this hypothesis, spelling errors from a small subset of the decoding data were examined. In the contraction conditions, there were six times as many errors involving vowels as consonants. The corresponding ratio was only 1 to 1 in the truncation conditions.

Relationship Between Decoding Performance and Rating Scores. In order to determine how well a participant's rating of a term-abbreviation pair correlated with his or her ability to decode the abbreviation, 2 × 2 contingency tables were created. A separate contingency table was constructed for each participant in condition (0) (thus only rule-generated abbreviations were considered). One axis of the table divided decoding responses into ''correct'' and ''incorrect.'' The other axis divided ratings (taken from the rating experiment) into ''high'' and ''low.'' A ''high'' rating was defined as above the mean rating given by that participant. For each abbreviation that a participant had to rate and decode, an entry was made in the contingency table. Contingency tables in which the marginal sum of any individual column or row was less than four were not examined, as the data were unsuitable for computing correlations.

Two correlations were computed for each participant: the phi coefficient and the tetrachoric coefficient. The latter coefficient has the advantage of being less sensitive to how a variable (i.e., high-low rating) is dichotomized. However, it also has a relatively high standard error (Guilford, 1950, pp. 332–339). From the individual correlation coefficients, a median was computed for each list condition. This was done separately for strict and liberal scoring and with the phi and tetrachoric coefficients. Not one of the medians for any condition exceeded 0.4. Thus, a participant's rating of an abbreviation is not a good indicator of whether he or she will decode it correctly.

Abbreviations Incorporating Endings. The ability to decode abbreviations incorporating endings (ending stimuli) was analyzed using strict scoring only. The results show that truncation abbreviations are correctly decoded more often than contraction abbreviations (.58 vs .46), $F(1,88)=6.04$, $p< .02$. As discussed above, this is probably due to the spelling assistance provided by the presence of vowels in truncation abbreviations. The data also show that in decoding abbreviations that incorporate endings, participants are much more likely to correctly produce the ending of a word (.52 correct in conditions 16 and 33) than when decoding abbreviations that do not incorporate endings (.12 correct in condition 0), $F(2,132)=52.21$, $p< .001$. Thus, the system for incorporating endings in abbreviations is effective during decoding, although it was detrimental during encoding. However, there was no significant effect due to abbreviation length.

Deviant Abbreviations (Liberal Scoring). Decoding performance on deviant abbreviations improved as the proportion of deviant abbreviations in the list increased, $F(1,88)=9.71$, $p< .01$. Thus, deviant abbreviations in lists (33) were correctly decoded 42% of the time while for lists (16), the score was 29% correct. The greater experience with processing deviant abbreviations in condition (33) probably accounts for this superiority in performance. There were no other significant effects.

Deviant Abbreviations (Strict Scoring). The effects reported for liberal scoring were also found for strict scoring. In addition, there was a significant interaction between the abbreviation technique used on the rule-generated abbreviations in the list and the proportion of deviant abbreviations, $F(1,88)=5.55$, $p< .05$. In condition (16), deviant abbreviations within contraction lists were easier to decode than those within truncation lists. The opposite was true for condition (33). The interpretation of this interaction is unclear, but it appears to be related to a similar interaction reported earlier for rule-generated abbreviations.

Rule Generated Versus Deviant Abbreviations. For each of the eight lists in conditions (16) and (33), rule-generated abbreviations were correctly decoded more often than were deviant abbreviations. This was true for both strict and liberal scoring. Using the binomial distribution, the probability of this occurring by chance is $p = .008$, two-tailed test. Thus, abbreviations formed by rules were easier to decode than abbreviations for which there were no rules.

Summary. When participants knew the abbreviation rules, there were no major differences in the ability to decode abbreviations generated by the truncation and contraction techniques. Truncation, however, was superior in the presence of abbreviations which did not follow a rule. Also, the spelling of a word decoded from a truncation abbreviation was more likely to be correct than one decoded from a contraction abbreviation. As for abbreviation length, variable-length abbreviations were decoded more often than fixed-length ones. But the former were also longer, and that probably accounts for the difference. With regard to the method used for incorporating endings into abbreviations, it did result in making those endings easier to decode. Finally, rule-generated abbreviations were decoded more often than were deviant abbreviations.

7. GENERAL DISCUSSION

7.1 Teaching Operators the Rules

These experiments studied abbreviation performance when participants knew the abbreviation rules. This can be contrasted to the Moses et al. (1980) experiments in which participants had no knowledge of the abbreviation rules. For encoding, the best performance that they found was 62% correct for truncation-variable length abbreviations after six exposures to the stimuli. In comparison, the present experiment found 81% correct with the same abbreviation technique, and an even better 92% correct for truncation-fixed length abbreviations (Moses et al., 1980 did not test the latter technique). This result confirms that abbreviations formed by a simple rule are easier to encode when operators know the rule.

The advantage of knowing the rule can also be seen when comparing rule generated to non-systematically generated (deviant) abbreviations. In the Moses et al.(1980) experiments, truncation-variable length abbreviations were no better than Army abbreviations, which are non-systematic. The same two sets of abbreviations were tested in the present experiments where participants knew the rules. Here, the truncation-variable length abbreviations were easier to produce than the non-systematic ones. Thus, the improvement in performance is due to the participant's knowledge of the rules and not to orthographic or other differences in the stimuli themselves.

However, knowledge of the abbreviation rules did not facilitate decoding. In the present decoding experiment, performance on both truncation-variable length and contraction-variable length abbreviations was about 66% correct. In the Moses et al. (1980) experiment, where participants did not know the rules, participants also performed equally well on these technqiues, with performance ranging from 70% correct after one exposure to the stimuli to 88% correct after six exposures. This better performance might be due to the fact that the Moses et al. (1980) variable-length rule created longer abbreviations on the average. In addition, the study phase for the two experiments were different as were some other factors.

There was one important decoding difference in the results reported by the two experiments. Moses et al. (1980) found no differences in decoding performance for rule-generated and non-systematically generated abbreviations. However, the present experiment found deviant abbreviations to be at a distinct disadvantage. Thus, the participant's knowledge of rules might interfere with his or her ability to decode abbreviations which do not follow the rules.

7.2 Choosing an Abbreviation Rule

The present experiments found truncation to be the better abbreviation technique. This was quite clear in the encoding task and marginally true in the decoding task. But the superiority of truncation over contraction might be even greater outside the laboratory. In the above experiments, participants had ample time and no distractions. Also, they saw each word instead of hearing it or mentally conceiving it. But most working environments are not so favorable. Contraction is a relatively difficult rule to apply as compared to truncation. If operators had to apply it to words not written on a piece of paper, or when under heavy stress, then the difference in performance between truncation and contraction would probably be even greater. However, as operators became more experienced, this difference would diminish.

Likewise, other experimenters have reported truncation abbreviations to be equal to or superior to abbreviations formed by other techniques. Moses et al. (1980) and Moses and Potash (1979) report this for both encoding and decoding for participants that did not know the rules. Likewise, Schneider, Hirsh-Pasek, and Nudelman (1981) report the same for encoding but not decoding. Using response-time measures, Rogers and Moeller (1981) found that with practice, truncation abbreviations are decoded faster than non-systematically generated (Navy) abbreviations.

Despite the apparent superiority of truncation, it may have disadvantages that prevent it from being the best choice for an abbreviation method. This issue revolves around the question of how many ambiguous abbreviations are produced by the truncation technique (Ehrenreich, in preparation). If there are many ambiguous abbreviations, then operators would be forced to learn numerous exceptions to the rule. The matter is further discussed below.

With regard to abbreviation length, the present experiments found no encoding differences for fixed- and variable-length abbreviations. However, this finding is probably not valid outside the laboratory. In the above experiments, participants saw the words written out, and this might not be the case in a working environment. Since the variable-length abbreviation method requires that an individual correctly count the number of letters in a word, it would probably lead to more errors than the fixed-length method. It thus seems wiser to use fixed-length abbreviations, particularly in situations where users are taught the abbreviation rules, and where encoding performance is important.

The last issue examined in these experiments was the ability to incorporate endings (ING, ED and S) into abbreviations. This was done by adding either a G, D, or S, as appropriate, to the end of an abbreviation. Performance with this technique was mixed. Although participants were more apt to correctly represent the ending when they decoded abbreviations, they were also more likely to err when encoding a word which had an ending. Unless it is crucial to the understanding of the message, it seems wiser to not incorporate endings into abbreviations. In many instances, extra space will have to be reserved on the screen and in computer memory to allow for the few instances where incorporating endings are important. In addition, the incorporation of endings complicates the abbreviation rules that the users must learn. When there are only a few words for which endings are critical, it would be better if these words not be abbreviated.

7.3 Limitations of a Single Abbreviation Rule

A single rule sometimes generates the same abbreviation for different words. When this occurs, a way must be found to disambiguate the abbreviations. One solution is to apply a second rule. For example, the truncation-fixed length technique might be chosen as the primary abbreviation rule for a list of words which include PROVIDE and PROVOKE. But the abbreviations produced by these two words are identical, i.e., PROV. A second rule, e.g., contraction-fixed length, could then be used to abbreviate these few words.

But using two abbreviation rules places a memory load on operators. They must learn which word is abbreviated by the secondary rule. By default, all remaining words are abbreviated by the primary rule. If the number of words abbreviated by the secondary rule is small, then this would not be too difficult. In any case, the amount of learning that would be required by this technique is much less than having to learn the correct abbreviation for each individual word in the vocabulary.

One way to help the operator remember which abbreviations were formed by a secondary rule is to mark them. In the above experiments, an asterisk was placed before those abbreviations that were not generated by the primary rule. Likewise, computer software can be designed so that an asterisk appears before such abbreviations. Then, when users encounter abbreviations that they must decode, they will know by what rule it had been formed.

However, there is no practical method for marking words so as to indicate how they are to be encoded. When an individual is tasked with entering a statement into a computer, he or she must decide whether the primary or secondary abbreviation rule is appropriate. This decision can be made by looking up the words in a manual or by having previously learned the information. In the encoding experiment, the information was artificially affixed to the word by attaching an asterisk. Thus the experiment does not really tell us how successfully operators can learn to use a primary and secondary rule when encoding words.

The problem created by using a secondary rule can be minimized by judiciously avoiding words which have identical abbreviations under the primary rule. Although it may not be possible to eliminate all such words, their number can be kept small by replacing them with synonyms. This is not to suggest that words strongly embedded in the operator's job-related vocabulary should be arbitrarily changed. But insofar as a vocabulary is being established for use on the system, some leeway in the choice of words is bound to be present. This would minimize the amount of learning required of the operator.

Looking back at the high performance observed in the encoding experiment, one can see that it is an overestimate of what to expect in a realistic situation. Since deviant abbreviations were marked with an asterisk, the observed encoding performance represents at best what might be expected if an individual had perfect knowledge of which words were to be abbreviated using the simple rule. For less skilled users, this would not be true and encoding performance would be lower (further discussion of this matter can be found in Ehrenreich, in preparation). The same statement is not true for the decoding performance observed in the experiment. It is legitimate to assume that the system can be programmed to show deviant abbreviations with an asterisk or other marking.

There are other solutions to the problem of a single rule generating identical abbreviations for different words. One possibility is for the operator to type only as much of the initial portion of the word as is needed to uniquely identify it (i.e., the minimum-to-distinguish technique described previously). For example, if TRANSPORT were the only word beginning with a T, then the operator need only type T in order to enter the word. However, if TRANSLATE were also in the computer's lexicon, then as a minimum, TRANSP would need to be entered in order to uniquely identify the word. This is equivalent to a truncation-variable length rule but with the length depending on the orthographic uniqueness of a word. But the minimum-to-distinguish technique also places a memory load on the operator who has to remember how many letters must be entered for each word. Thus, this technique has the same drawback as the technique of using a primary and secondary rule. The more words being abbreviated, the greater the likelihood that some will have their first few letters in common. For both abbreviation systems, this complicates their use. The question then is, Which one results in faster learning and longer retention by the operator? Further comparison of the above two techniques can be found in Ehrenreich (in preparation).

Another possible solution to the problem of ambiguous abbreviations is to assure that they always occur in mutually exclusive modes of operation.[4] For ex-

---

[4] This consideration was brought to my attention by Robert Solick.

ample, the abbreviation C can represent either the COPY command or the CHANGE command. But since COPY might occur in the job-control mode while CHANGE occurs only in the text-edit mode, this sameness of abbreviation presents no practical problem. Likewise, even in the same operating mode, it is possible for two words to have identical abbreviations and yet be so syntactically and semantically disimilar that they can be disambiguated by the contexts in which they occur. In a query-language situation, where the number and variety of permissible sentences is large, programming such a capability may involve a greater investment in software than is warranted by the problem. But the situation may be different when the operator is filling-in-the-blanks on a "form" being displayed on the VDT. Here, the exchange between operator and computer is restricted, and words with identical abbreviations might only occur as responses to different questions. In such cases, the system can easily identify the correct meaning of an abbreviation since the alternative is not a permissible option. Thus, identical abbreviations may be acceptable, eliminating the need for a secondary abbreviation rule.

7.4 Guidelines

Based upon the preceding discussion, a set of guidelines are presented in Table 3. These guidelines are an extrapolation, in practical terms, of the results from these and other experiments on abbreviations.

The basic premise of these guidelines is that operators can work with abbreviations more effectively if they understand how they were created. One way to effect this knowledge is to generate abbreviations via a primary and secondary rule. The primary rule is used to abbreviate the great majority of words and the secondary rule is used as a backup. Many forms of primary and secondary rules

TABLE 3
Guidelines for Generating Abbreviations

1. A *simple,* primary rule should be used to generate abbreviations for most items and a *simple,* secondary rule used for those items where there is a conflict.

2. Abbreviations generated by the secondary rule should have a marker (e.g., an asterisk) incorporated into them.

3. The number of words abbreviated by the secondary rule should be kept to a minimum.

4. Operators should be familiar with the rules used to generate abbreviations.

5. Truncation is an easy rule for operators to work with but it may also produce a large number of identical abbreviations for different words.

6. Fixed length abbreviations are preferable to variable length ones.

7. Abbreviations should not be designed to incorporate endings (e.g., ING, ED, S).

8. Unless there is a critical space problem, abbreviations should not be used in messages generated by the computer and read by the operator.

can be suggested; for example,truncation-fixed length as a primary rule and contraction-fixed length as a secondary rule. In fact, the minimum-to-distinguish rule is simply a primary rule with a reiterative secondary rule, i.e., primary rule—use the first letter of a word to abbreviate it; secondary rule—if the abbreviation is not unique, add another letter (repeat secondary rule as necessary).

Although the question of which primary and secondary rule is best has not been answered, three criteria are obvious. First, the rules must be easy for an operator to understand. Second, they must be simple to apply, preferably without the aid of paper and pencil. And finally, the primary rule should be able to uniquely abbreviate all but a few of the words. Any reasonable choice for a pair of rules, even if they are not the "best," is sure to produce encoding performance that is superior to the performance obtained by using non-systematic abbreviations.

Although it is also desirable to improve decoding performance, none of the techniques tested here had any effect. Fortunately, decoding appears to be intrinsically easy, with a learning curve that reaches a high level after only a few trials. As reported earlier, participants in the Moses et al. (1980) experiments were able to correctly decode the stimuli 70% of the time after one exposure, and 88% of the time after six exposures.

The guidelines presented here cannot stand alone and a human-factors specialist or system designer must still consider the individual system and its operators when adopting them. But hopefully this paper and its guidelines provide a body of information and a set of options that will help in developing "user-oriented" abbreviations for automated systems.

8. ACKNOWLEDGEMENTS

The authors wish to express their appreciation to a number of people who contributed to this paper: Frank Moses and John Mellinger for valuable discussions held with them, and Gail Rowan and Andrew Zbikowski for performing the statistical analyses.

9. REFERENCES

Ackroff, J.M., & Streeter, L.A. *Memorability of abbreviations and their referents*. Paper presented at the 52nd Annual Meeting of the Eastern Psychological Association, New York City, April 1981.

Ehrenreich, S.L. Query languages: Design recommendations derived from the human factors literature. *Human Factors*, 1981, *23*, 709–726.

Ehrenreich, S.L. *Designing abbreviations for automated system: Alternative heuristics*. Manuscript in preparation, 1982.

Ehrenreich, S.L., & Porcu, T. *Abbreviations: Improving operator performance on battlefield automated systems.* (Technical Report 556). Alexandria, VA: US Army Research Institute for the Behavioral and Social Sciences, in press.

Guilford, J.P. *Fundamental statistics in psychology and education* (2nd ed.). New York: McGraw-Hill, 1950.

Hodge, M.H., & Pennington, F.M. Some studies of word abbreviation behavior. *Journal of Experimental Psychology,* 1973, *98* (2), 350–361.

McBride, D.K., Lambert, J.V., & Lane, N.E. *The development of a standardized abbreviation algorithm (ABBREV).* Paper presented at the 52nd Annual Meeting of the Eastern Psychological Association, New York City, April 1981.

Moses, F.L., Mendez, C.H., & Ehrenreich, S.L. *The effect of learning and context on abbreviation intelligibility.* Paper presented at the 88th Annual Convention of the American Psychological Association, Montreal, September 1980.

Moses, F.L., & Potash, L.M. *Assessment of abbreviation methods for automated tactical systems.* (NTIS No. AD A077 840, Technical Report 398), Alexandria, VA: US Army Research Institute for the Behavioral and Social Sciences. August 1979.

Roger, W.H., & Moeller, G. *The effects of rules and redundancies on the processing of abbreviations* (NSMR Report 958). Groton, CT; Naval Submarine Medical Research Laboratory, November 1981.

Schneider, M.L., Hirsh-Pasek, K., & Nudelman, S. *Abbreviations in limited lexicons.* Paper presented at the 52nd Annual Meeting of the Eastern Psychological Association, New York City, April 1981.

Streeter, L.A., Ackroff, J.M., & Taylor, G.A. *On abbreviating command names.* Paper presented at the 52nd Annual Meeting of the Midwest Psychological Association, St. Louis, May 1980.

7

Models for the Design of Static Software User Assistance

M. L. SCHNEIDER

Sperry Univac
Blue Bell, PA. 19424

This paper proposes a model of user sophistication composed of five levels: Parrot, Novice, Intermediate, Expert, and Master. These levels can be defined in terms of the assumed chunk size employed by the user. A model of task processing decomposes a transaction into five stages: Task Analysis, Semantic Analysis, Syntactic Analysis, System Performance, and Response Analysis. The relationship between these two models provides possible guidelines for determining the information presented to users.

1. INTRODUCTION

One of the "axioms" for ease-of-use is: "Help systems are necessary" (Clark, 1980). User assistance information can be classified into two groups: dynamic, providing information about the current state of the system, and static, providing information that is written in advance. The information contents of static assistance are examined in this paper. While an increasing number of software systems provide some form of user assistance (Relles, 1979), the information is usually provided without regard to its usage. In general, assistance is nothing more than an "electric reference manual." When the information is factored into smaller entities, a layered approach is generally used; the user can request additional details about a specific topic. This addresses the problem of verbosity, but only indirectly considers the expertise of the requester.

This paper proposes cognitive factors that may impart more specific information factoring: different levels of user sophistication (the User Taxonomy) and different segments of task performance (the Transaction Taxonomy). The interaction between these two taxonomies can provide guidelines for improved static information presentation.

137

2. USER SOPHISTICATION TAXONOMY

The developmental levels of computer interface acquisition proposed by this taxonomy are:

1. Parrot
2. Novice
3. Intermediate
4. Expert
5. Master

Each level is characterized by the chunk size assumed to be employed by the user, language scope, and the degree of generalization or abstraction of concepts. As the user progresses through the levels in the model, the assumed chunk size increases from characters (Parrot) to procedures (Expert); the language scope ranges from limited functions (Novice) to full functional (Expert); and, the degree of generalization begins with concrete (Novice and Intermediate) and ends with generative (Master). The change in system knowledge is manifested through an increased competence in the commands that are regularly used and an awareness of additional functions available within the system or language.

This taxonomy arises from qualitative observations of computer usage in a wide variety of software systems. Although the underlying concepts are largely based on concepts of complex task acquisition, it has not been fully tested. This taxonomy describes an individual's expertise or sophistication in a single software system or language (or subset thereof) and may not be transferable. A person might be an Expert in one system and a Parrot in another.

The level at which an individual stops progressing appears to depend upon the learning process and the demands placed upon the person by the tasks.

2.1 The Parrot

An individual at the lowest level in the taxonomy, the Parrot, has minimal knowledge of the computer system. The Parrot operates with the smallest chunk available: a single character. A user at this level approaches the computer system and types strings of characters. This individual does not think, question, understand, or synthesize the text entered. Commands, or sequence of commands in some cases, corresponding to the text may be moderately complex. Yet, the user is unaware of this complexity. Satisfaction is derived simply by having the computer perform the task.

When the qustion ''What am I doing?'' is asked, the Parrot is ready to progress to the next stage of sophistication: the Novice.

2.2 The Novice

With experience, a user begins to understand several isolated concepts and is able to choose a specific lexical entry (command) for a function. The user now attaches meaning to the input; the information is specific but not complex. Semantically, the items are considered in the concrete, not in the abstract. The Novice may ask, ''What does this command item do?'' not ''What can it do?'' By now, the user has a minimal command of the grammar, but usually operates on an item-by-item basis; an item is postulated as the chunk size. For example, the Novice may recognize a verb and one or more objects in a command, even if the grammar allows modifiers in the verb phrase or in the object phrase.

Unlike the Parrot, the Novice analyzes each item, thus extracting lexical information. The language components now have meaning and can be used in a flexible manner.

2.3 The Intermediate

Whereas the Novice concentrates on items in isolation, the Intermediate operates with items in fields and with fields in statements. The Intermediate collects items into a larger chunk—the statement. The use of this larger chunk encourages syntactic and semantic conciseness in the grammar, allowing the user to minimize keystrokes.

At times, the Intermediate user may link statements into command ''chains'' such as compile . . . collect . . . execute. Even so, each command is still considered in isolation. The user generally waits until a function has been completed before proceeding to the next request, wishing to see the result of a command before continuing with the task.

The Intermediate begins to concentrate on the task, rather than its components. Use of the full language may be restricted by a lack of knowledge. Thus, the Intermediate continues to expend significant effort on language details.

At this point in the user's development, the more subtle grammatical rules become evident. A Novice would use a default, unaware of the fact that an item can be specified. An Intermediate would consciously use a default in order to reduce keystrokes or save time. Initially, the Intermediate uses knowledge in a specific problem domain. Later, this information is generalized, allowing new problems to be solved.

Toward the end of the Intermediate level, considerable skill in the understanding and manipulation of a segment of the command set has been achieved. With the increased use of larger syntactic chunks, each requires less attention. This is the process of automatization. Thus, increased attention can be given to the entire task, rather than to the mechanisms required for its performance.

With further experience and increased task requirements, the Intermediate can evolve into an Expert, subordinating the computer language to the task.

2.4 The Expert

While the Intermediate attempts to solve problems via a series of isolated commands, the Expert realizes that an interconnected collection of statements can be more productive for certain tasks. The Expert, it is postulated, uses the largest chunk: a program or procedure. Because commands are now interrelated, the scope of the syntax and semantics expands. The syntactic elements are abstract rather than concrete. Data structures provide the vehicle for producing abstract objects. For example, a variable would be used to represent a file name or a string. The Expert continues to retain the command, together with other defined procedures, as language primitives.

Control structures are useful if the direction of flow between statements is to be modified. Using these structures requires a modification of the user's thought process. A Novice or Intermediate user may not foresee the success or failure status of a command as an object on which operations are defined. An Expert thinks about the possible outcomes of commands and has the ability to take appropriate action. While Novice and Intermediate users operate with concrete syntactic constructions, existing within a specific, restricted semantic scope, the Expert expands his language knowledge to cope with complex structures and abstractions.

Practically speaking, the Expert has the ability (though not necessarily the need) to accomplish any function within the system. The Expert is completely facile with the language and can deal with the language at the global ''metalinguistic'' level.

2.5 The Master

The Expert has the ability to use the language with relative ease. Since any computer language is restricted in scope, it can limit a user (for example, the inability to have abstract data types in FORTRAN 77). The Expert, knowing the scope of the language, is constrained when faced with a new problem whose solution cannot be derived from existing functions or objects within the system. The Master transforms this finite system into a generative one. When faced with the above situation, he creates, not derives, a new syntactic element within the system. Thus the Master expands the existing system, creating new objects and functions. Unlike the lower four levels, the chunk size is not a determining factor.

3. TRANSACTION TAXONOMY

While the sophistication level of the user is important, it is necessary to know how a transaction is processed in order to factor assistance information. A transaction in this paper is defined as the task contemplated by the user (for example, writing a program, ''checking in'' an airline passenger, or performing a data-base query).

The five-stage transaction taxonomy shown below builds upon a simple taxonomy (command and data input, processing, and system output) by expanding the first operation—input—into its semantic and syntactic components as suggested by Shneiderman (1979).

STAGE ACTION

I	Task Analysis
II	Semantic Analysis
III	Syntactic Analysis
IV	System Performance
V	Response Analysis

3.1 Stage I—Task Analysis

In the first stage, the user decomposes a single conceptual task into its component subtasks, determining the specific commands required for task completion. The user asks the question, ''What steps and commands are necessary to perform the overall task?'' For example, running a program (the single conceptual task) may require the following subtasks: editing, compilation, collection, and execution. It is possible that more than one step can be included within a single command (for example a compile-load-go) or more than one subtask is required within each subtask (for example, the editor commands).

The cognitive proceses at this stage may include some or all of the following steps:

1. Identification of the full task.
2. Decomposition of the task into its subtasks or steps.
3. Definition of the conceptual operation for each step.
4. Choice of the appropriate command for the implementation of each step.

It should not be assumed that all commands will be chosen at the outset. It is highly probable that an individual will determine the conceptual operation for the first subproblem, choose an appropriate command, perform it, assess the result, then progresss to the next conceptual operation, the choice of which may be influenced by the result of a previous task.

Once the conceptual operation has been defined, a user may wish to examine the set of commands for its implementation. It is possible to relate commands and conceptual operations in two ways: define a conceptual operation for commands that are conceptually related; or its antithesis, to extract from a conceptual operation its constituent commands. By iterating between these perspectives, it should be possible for the user to determine a command that allows the conceptual operation to be performed.

A command subset of a hypothetical editor illustrates this iterative approach. Consider the command "LOCATE" (this searches the text printing the lines whenever a string occurs). The specific to general relationship would be:

$$\text{"LOCATE"} \longrightarrow \textit{search}$$
$$\textit{print}$$

The general concept *print* may refer to a number of commands that, if successfully executed, print a line:

$$\textit{print} \longrightarrow \text{"PRINT"}$$
$$\text{"LOCATE"}$$
$$\text{"FIND"}$$
$$\text{"GOTO"}$$
$$\text{"NEXT"}$$

If all commands of the concept *search* print a line, then the structure could be represented as:

$$\textit{print} \longrightarrow \text{"PRINT"}$$
$$\text{"GOTO"}$$
$$\text{"NEXT"}$$
$$\textit{search} \longrightarrow \text{"LOCATE"}$$

A similar grouping can occur for "GOTO" and "NEXT." When explanations are provided (basic semantic information) within the above framework, the user can obtain the information in a unified manner.

3.2 Stage II—Semantic Analysis

In the second stage, the scope of the command is considered by the user. Upon entry to the semantic analysis, the command is conceptual in the broadest sense. Now it must be refined into its detailed semantic components.

The question: "What do I want to do?" is asked by the user. The user must be cognizant of two semantic concepts: definition of the data and the control of the process. A sorting program illustrates the type of information considered by the user. A user must be aware of the data restrictions (for example, numerics only, alphanumerics, maximum number of items, maximum number of fields, etc.) and the method(s) of data storage or entry. In addition, information is required to control the processing (ascending, descending, key(s), collating sequence, etc.). At the semantic stage, it is unnecessary to know how to encode this information.

3.3 Stage III—Syntactic Analysis

When a user reaches the third stage, encoding the information, the correct function has been chosen and the semantics for task completion are understood. Now the question is ''How do I do it?'' The translation of the conceptual operation into the input format is purely mechanical. The user requires syntactic information and techniques that facilitate this transformation. The form of the human-computer interface (command language, dialogues, menus, function keys, etc.) has a primary impact at this stage.

3.4 Stage IV—System Performance

System response, the fourth stage, can be treated as a ''black box.'' The underlying architecture that supports the interface is outside the scope of this paper.

3.5 Stage V—Response Analysis

The analysis and interpretation of the response produced by the software is the final stage of a transaction. The user now asks, ''What have I done?'' The primary goal of a response is to provide the user with relevant information. Unnecessary details that obscure this information should be avoided. Two independent topics should be considered: verbosity and information content (Schneider, 1980).

For example, if the task is to assign the file, MYFILE, there are a number of possible responses if it is successful (ordered by increasing verbosity and content):

1. > {a prompt for the next command}
2. READY {, OK, COMPLETE, . . . }
3. File MYFILE has been assigned.
4. MYFILE assigned with the PUBLIC, and CATALOG options.
5. File MYFILE has been assigned. It can be used by anyone (PUBLIC) and will exist for one day (CATALOGUED) unless otherwise requested. To keep the file longer than one day contact the file administrator.

The last response is an example of layering. Three items of information have been displayed:

1. The name of the assigned file.
2. The file attributes.
3. The administrative procedure required to keep the file.

In a similar manner, it is possible to design a layered HELP function (a user-initiated request for assistance).

A command may not always teminate successfully. Useful and meaningful error messages are important. They should reflect the user's concept of the task, not the implementer's view. Good error reporting should provide sufficient information for the user to:

1. Understand the nature of the error;
2. Understand the source of the error;
3. Understand the methods for recovery or correction.

Again the questions of verbosity and information content are important. Verbosity may be correlated with the number of times an individual has seen the message, while information content should be related to the levels in the user taxonomy and task requirements.

4. INTERACTIONS BETWEEN TAXONOMIES

The user and transaction taxonomies should not be considered in isolation. Based upon the sophistication level of the user, the scope of assistance may vary. Different segments of the transaction taxonomy need to be emphasized or deemphasized. The method of assistance presentation provided to individuals at different sophistication levels for the same transaction may differ. In order to better understand the type of assistance applicable at each level of use, it is necessary to examine the requirements of users at each sophistication level.

4.1 Parrot

A parrot operates in a simple "transcription mode." There is no consideration of input variability. The best form of input assistance is an example or a single choice from a single-level menu system. The latter is analogous to function keys. By careful design, either of these approaches can be extended to assistance forms suitable for a Novice.

Only two basic responses can exist for the Parrot: the function completed successfully, or it was unsuccessful. If an unsuccessful response is provided, it can only state that the command was incorrectly entered and should be entered again (a Parrot does not comprehend the command's contents).

If the system is unable to perform the task at this time, it can be suggested that the user try later. Since task completion is the reward for successful command entry, this information should always be provided to the user. Thus, at the Parrot level there is only one type of input assistance: an example.

4.2 Novice

The novice may not distinguish between the first three stages of a transaction (Task, Semantic, and Syntactic Analysis). Thus, these stages should not be differentiated if the user's perspective is to be reflected in the interface. The system should lead the user from the determination of the subtask(s), through the isolation of the correct command and the determination of its semantic components, to the encoding of the information.

Once the user is ready to provide data for the command, a number of techniques can be applied. As stated earlier, continuity between the first three stages is important; the user should be unaware of any distinct phase of the transaction. Since the traditional command format may be inapplicable to the Novice, menus could be used for Stages I and II followed by a mixture of menus, dialogs, and ''fill-in-the-blanks'' for Stage III.

This expands the syntactic assistance to two levels:

Assistance Type	Sophistication Level
Example	Parrot
Command Determination and The Simplest Command Form	Novice

Irrespective of the technique, the computer should take the initiative; the Novice may not know what information is required, or even if it is available. Thus, it is incumbent upon the assistance system to announce its existence. Information for clarification, however, should be provided only upon demand. To do so automatically may unnecessarily confuse or annoy the user.

Responses, aside from providing information to the user, should indicate the successful completion of the command in a non-null form (something more than a prompt). A Novice, lacking confidence in the ability to control the system, may require this positive reinforcement.

4.3 Intermediate

Because the Intermediate is familiar with the system, the user, not the computer, should take the initiative. An individual at this sophistication level has the ability to decompose a task into its subtasks and determine an appropriate command (Stage I). Since the components of the system are known to exist, even if not understood, information should be factored into the following topics: command semantics, command syntax, and field or keyword semantics and syntax. Since individuals generally employ a subset of commands in a system (Huckle, 1980), Novice-type assistance is still required for those used less commonly.

Assistance in the semantic and syntax analyses (Stages II and III) require additional information. As a user gains experience with a command, defaults are better understood, overridden, or modified. Thus, the scope of the command perceived by the user is extended. The semantic and syntactic expansion of commands requires that two new levels of assistance must be added:

1. The most common form of the command. This will occur when some commonly defaulted items are overridden.
2. The command is used in its full form. This occurs when no item is defaulted.

Thus, the number of levels are increased to four:

Assistance Type	Sophistication Level
Example	Parrot
Command Determination and	
The Simplest Command Form	Novice
The Common Command Form	Intermediate
The Full Concrete Command Form	Intermediate

When the semantics and syntax of a command are not complicated, two or even one of the above forms may fulfill the information requirements.

Because the Intermediate operates in a terse mode, abbreviated forms of the command should be provided. This includes not only contracted forms of the strings within the command (name, keywords, flags, etc.), but the items that can be defaulted and the values supplied.

The layered approach for responses should be available. As in the case of information required for the input of a command, the user should be able to request specific information. The advantages (terseness and specificity) of requesting specific information are offset by the need for a query language with its increased complexity.

4.4 Expert

The execution of a procedure is similar to a batch system; a number of commands are executed without the user's direct intervention. Thus, the needs of the Expert differ from the Intermediate in three ways.

1. The transaction stages considered prior to entering a command require a different emphasis because data and control structures are now a part of the user's command repertoire.
2. There is a need for assistance in the monitoring of an executing command, since they are executed in a "batch environment."
3. A different type of response structure is needed, since it must be interpreted directly by a command within the software without human intevention.

Within the first two stages, an increase in the type of information exists, reflecting the added control and data structures employed by the Expert. These new structures may be implemented within an existing command or via new commands. Assistance and instruction in the methods of building macros, procedures, and programs are useful for the Expert. These new functional elements are reflected not only in Stages II and III, but their concepts must be included in Stage I.

Control and data structures are now used in the development of procedures. This places additional demands upon the response segment. Whereas in the lower sophistication level interfaces, the responses must be understood by a human; in a procedure, responses must be understood by the software. The abstract nature of the command requires additional syntactic information. When a command has constructs that relate only to these structures, they must exist only in the information supplied to the Expert. Thus, in addition to the three assistance levels applicable to the Novice and Intermediate users, a fourth level, containing the expanded language view must be included. The five levels of assistance are shown below:

Assistance Type	Sophistication Level
Example	Parrot
Command Determination and the Simplest Command Form	Novice
The Common Command Form	Intermediate
The Full Concrete Command Form	Intermediate
Control and Data Structures with the Full Command Form	Expert

5. CONCLUSION

On a theoretical basis, it is possible to factor software user-assistance information into three independent categories:

1. verbosity
2. user sophistication
3. task segmentation

Although it is possible to prepare guidelines for the further classification of information within each category, only experimental investigations will validate these suppositions. Currently, studies of specific topics and the testing of the two models are in progress.

6. ACKNOWLEDGEMENT

I would like to thank K. Hirsh-Pasek with whom I have had many loud and fruitful discussions.

7. REFERENCES

Clark, I. A. *How to "help" help* (IBM Report HF022). Hursley Park: IBM United Kingdom Laboratories Ltd., 1980.

Huckle, B. A. Designing a command language for inexperienced users. In D. Beech (Ed.), *Command language directions*. Amsterdam: North-Holland Publishing Company, 1980, 199–212.

Relles, N. *The design and implementation of user-oriented systems*. Ph.D. Thesis, University of Wisconsin, 1979.

Schneider, M. L., Wexelblat, R. L., & Jende, M. S. Designing control languages from the user's perspective. In D. Beech (Ed.), *Command language directions*. Amsterdam: North-Holland Publishing Company, 1980, 181–198.

Shneiderman, B., & Mayer, R. Syntactic/semantic interactions in programmer behaviour: A model and experimental results. *International Journal of Computer and Information Sciences*, 1979, *7*, 219—239.

8

A Human Factors Engineer's Introduction to Speech Synthesizers

*PAUL ROLLER MICHAELIS
AND RICHARD H. WIGGINS*

Texas Instruments Incorporated
MS 238, P.O. Box 226015
Dallas, Texas 75266

The first half of this chapter describes how speech synthesizers work. Most speech synthesizers are based on mathematical models of the human vocal tract. Commonly used techniques for modeling the human speech mechanism include channel synthesis, formant synthesis, and synthesis based on linear predictive coding. Of greater importance to the designer of human-computer interfaces is the fact that synthesizers also differ in the format of their input data. The input data for analysis/synthesis systems are the parameters that control the speech production model, which were previously obtained by analyzing a digital recording of a person saying that word or phrase. Another type of system accepts phonemes or even text as input, and does what is called synthesis-by-rule: the parameters that control the speech production model are computed at the time of synthesis by applying pronunciation rules and dictionaries to the input data.

The second half of this chapter is devoted to human factors concerns. Following a discussion of when speech synthesizers should be used, this section addresses the two major limitations of speech synthesizers: voice quality and vocabulary restrictions. Various techniques are described for improving the intelligibility and aesthetic qualities of synthesized speech. The problem of vocabulary restriction is addressed by providing a list of about 200 general words useful for all applications, and describing a technique for composing the supplementary application-specific word lists. The chapter closes by examining trends and recent advances in speech synthesis technology.

1. INTRODUCTION

The most common type of auditory communication from machines to people is the tonal output of such devices as buzzers, bells, horns, or whistles. Only in extraordinary cases has the human factors engineer had the luxury of designing an interface in which the machine actually talks to the operator. Examples of speech-output interfaces can, however, be found in several military aircraft, where specific prerecorded messages are played to warn the pilot of certain abnormal conditions. There is no doubt that from a human factors point of view, the higher bandwidth of speech output makes it superior to tonal signaling systems for many applications. If that is the case, why aren't speech-output systems more common?

Speech-output human-machine interfaces have been rarely used because they have failed to satisfy three basic requirements: reliability, cost, and voice quality. Tape and other analog playback mechanisms usually fail to satisfy the first requirement because they have moving parts that wear out. One way to circumvent that difficulty and obtain a reliable speech-output system is to make the system completely solid state.

A straightforward method for achieving voice output with solid-state devices is to encode the speech waveform digitally. The process of converting an analog waveform to digital sequences is well understood: sample the amplitude of the analog signal at a rate that is twice the highest frequency component of the signal. Thus, to represent a speech signal of up to four kilohertz (i.e., 4,000 cycles per second), the waveform is first filtered to remove frequencies above four kilohertz and the amplitude is then sampled 8,000 times per second. These amplitude samples may be stored in a solid-state memory. To regenerate the analog signal, a standard digital-to-analog converter is used followed by a smoothing or low-pass filter. However, even for voice response systems with medium-sized vocabularies, this technique can require excessive amounts of storage. The amplitude samples are usually represented with 12-bit accuracy, and when sampled 8,000 times per second, the resultant data rate is 96,000 bits per second. At these rates, 15 typical English words would require about one million bits of memory. Modest memory savings can be achieved by nonlinear encoding schemes that represent small amplitudes with greater accuracy than large amplitudes. Such methods achieve equivalent voice quality with as few as eight bits per sample, but memory requirements still remain excessive. In general, although these digital recordings satisfy reliability and voice-quality requirements, the large memories required often make them too costly.

As a result of these technical considerations, both analog and direct digital playback systems have proven unsatisfactory for most applications. However, the designers of speech-output interfaces are not restricted to playback systems. They may also use devices that synthesize speech by electronically simulating the human vocal mechanism. Rather than digitize speech by encoding the waveform, the user of speech output may use a speech synthesizer and merely encode the parameters that control the speech production model. By the use of a simplified model whose characteristics change very slowly, large reductions in the required storage space (by factors of 50 to 200) may be obtained.

Although electronic speech synthesizers have been available for many years, these, too, were rarely used by interface designers because of their high cost and, at times, the marginal quality of their voices. Today, however, advances in highly reliable semiconductor technologies are greatly reducing the cost of complex digital processing, while simultaneous advances in speech synthesis techniques are offering dramatic improvements in voice quality. The result is that reliable, inexpensive, high quality speech output is now available to the human factors engineer.

Two major sections follow. The first describes how modern electronic speech synthesizers work. This section stresses aspects of their operation with which human factors engineers should be familiar, because the mechanisms by which these devices produce speech also create a new set of problems. The second section addresses and provides solutions to these problems.

2. FUNDAMENTALS OF SPEECH SYNTHESIS

Most speech synthesizers are electronic models of the human vocal tract. So to understand how synthesizers produce speech, it is first necessary to understand how humans produce speech.

2.1 Human Speech Production

The human speech mechanism is complex, and this section presents a comparatively simple model of it. (See Ladefoged, 1975, for a thorough description.) The major physical components of the human speech mechanism include the lungs, the vocal cords, and the vocal cavity, as shown in Figure 1.

In the generation of speech sounds, air is forced from the lungs past the vocal cords and through the vocal cavity. The pressure with which the air is exhaled determines the final amplitude, or "loudness," of each speech sound. The action of the vocal cords on the breath stream determines whether the resultant speech sounds will be voiced or unvoiced. The voiced sounds of speech (for example, the "v" sound in the word "voice") are produced by tensing the vocal cords while air is forced from the lungs. The tensed vocal cords interrupt the flow of air, resulting in the release of air in short periodic bursts. The frequency with which these bursts are released imparts pitch to the voice; the greater the frequency with which the bursts are released, the higher the pitch.

Unvoiced sounds (for example, the final "s" sound in "voice") are produced when air is forced past relaxed vocal cords that do not periodically interrupt the air flow. The sound is generated by audible turbulence in the vocal tract. A simple demonstration of the role of the vocal cords in producing voiced and unvoiced sounds can be had by placing one's fingers lightly on one's larynx, or voice box, while slowly saying the word "voice"; the vocal cords will be felt to vibrate

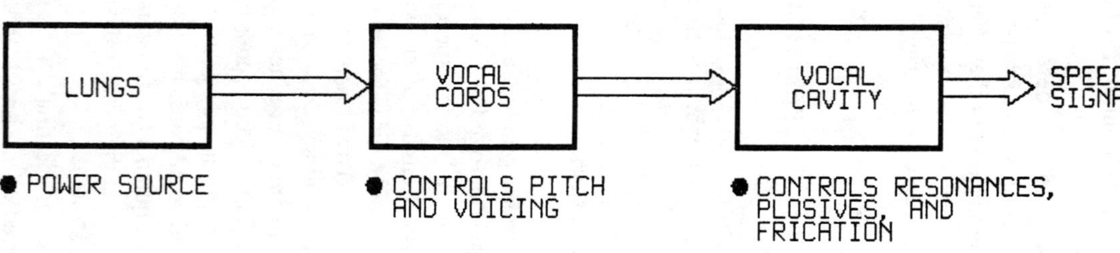

FIGURE 1. The major components of the human vocal mechanism.

for the ''v'' sound and for the double vowel (or diphthong) ''oi'' but not for the final ''s'' sound.

The sound-generating mechanisms described above produce what is called the excitation signal for speech. There are only three variable parameters in the excitation signal: its amplitude, the proportion of it that is voiced or unvoiced, and, if it is voiced, its fundamental pitch. This can be easily demonstrated. If one were to hold one's mouth wide open, without any movement of the jaw, tongue, and lips, the only remaining changeable characteristics of sound generated by the vocal system would be the above three parameters.

At any given time, the excitation signal will actually contain sounds at many different frequencies. A voiced excitation signal is periodic; the energy in its frequency spectrum lies at multiples of the fundamental pitch, which is equal to the frequency with which the vocal cords are vibrating. An unvoiced excitation signal contains a random mixture of frequencies similar to what is generally called white noise.

The vocal cavity ''shapes'' the excitation signal into recognizable speech sounds by attenuating certain specific frequencies in the excitation signal while amplifying others. The vocal cavity is able to accomplish this spectral shaping because it resonates at frequencies that vary depending on the position of the jaw, tongue, and lips. Frequencies in the excitation signal are suppressed if they are not near a vocal cavity resonance. However, vocal cavity resonances tend to amplify, or make louder, sounds of the same frequency in the excitation signal. The resulting spectral peaks in the speech sounds are called formants. Typically, only the three or four lowest-frequency formants will be below 5,000 hertz. These are the formants that are most important for intelligibility.

The sounds of human speech can be variously categorized according to the place of articulation, manner of formation, voicing, and the like. For spoken English, a simplified breakdown according to manner of formation would include the following four categories: vowel, nasal, fricative, and plosive sounds.

TABLE 1
Examples of of English Speech Sounds,
Classified According to Their Manner of Formation

	Voiced	Unvoiced
Vowel	E as in ''Easy'' A as in ''Answer''	(Not found in spoken English)
Nasal	M as in ''Man'' N as in ''Nasal''	(Not found in spoken English)
Fricative	V as in ''Vast'' Z as in ''Zoo''	F as in ''Fast'' S as in ''Sue''
Plosive	B as in ''Bat'' D as in ''Den''	P as in ''Pat'' T as in ''Ten''

In the formation of vowels, such as the ''e'' sound in ''speech'' and the diphthong ''oi'' in ''voice,'' the breath stream passes relatively unhindered through the pharynx and the open mouth. In nasal sounds, such as the ''m'' and ''n'' in ''man,'' the breath stream passes through the nose. Fricative sounds are produced by forcing air from the lungs through a constriction in the vocal tract so that audible turbulence results. Examples of fricatives include the ''s'' and ''ch'' sounds in ''speech.'' Plosive sounds are created when the vocal cavity is completely closed by the lips or tongue and the air pressure built up behind the closure is then suddenly released. The word ''talk'' contains the plosive sounds ''t'' and ''k.'' Except when whispering, the vowel and nasal sounds of spoken English are voiced. Fricative and plosive sounds, however, may be voiced (as in ''vast'' or ''den'') or unvoiced (as in ''fast'' or ''ten'').

Analogies of the human vocal mechanism can be found in those musical instruments that produce sounds by passing an excitation signal through controlled, variable resonators. In the case of trumpets, trombones, and tubas an excitation signal generated by rapid vibrations of the lips rather than the vocal cords can be shaped into different musical notes, or even different sounds, by changing the size and number of the resonating ''cavities'' through which this excitation signal passes.

2.2 Speech Production Model

Figure 2 shows a simplified flow diagram of a typical speech synthesizer. This synthesizer has two major components: the source function model that produces the excitation signal, and the model of vocal tract resonant characteristics.

The component that produces the excitation signal has two signal generators. One generates a periodic signal that simulates the sound produced by vibrating human vocal cords. The other produces a random signal that is suitable for modeling unvoiced sounds. Thus, when a synthesizer needs to generate a voiced sound, such as the ''e'' in ''speech,'' it uses the periodic output from the first signal generator; for the unvoiced ''sp'' and ''ch'' sounds in ''speech,'' the synthesizer would instead use the random output from the other signal generator. In some synthesizers, a weighted combination of the random and periodic excitation is used. This can be useful in generating voiced fricative sounds (for example, the ''z'' sound in the word ''zoo''). However, most synthesizers restrict the excitation source so that it is entirely modeled by either the voiced or unvoiced excitation. Alternation of the excitation can be controlled by a two-valued voicing parameter, usually referred to as the voiced/unvoiced decision.

Next, the source function is scaled by an energy or amplitude parameter, which allows the synthesizer to control the loudness. Finally, if the synthesizer is to generate something other than a monotone, it is necessary for the period of the voiced excitation function to be variable. The parameter that controls this is called

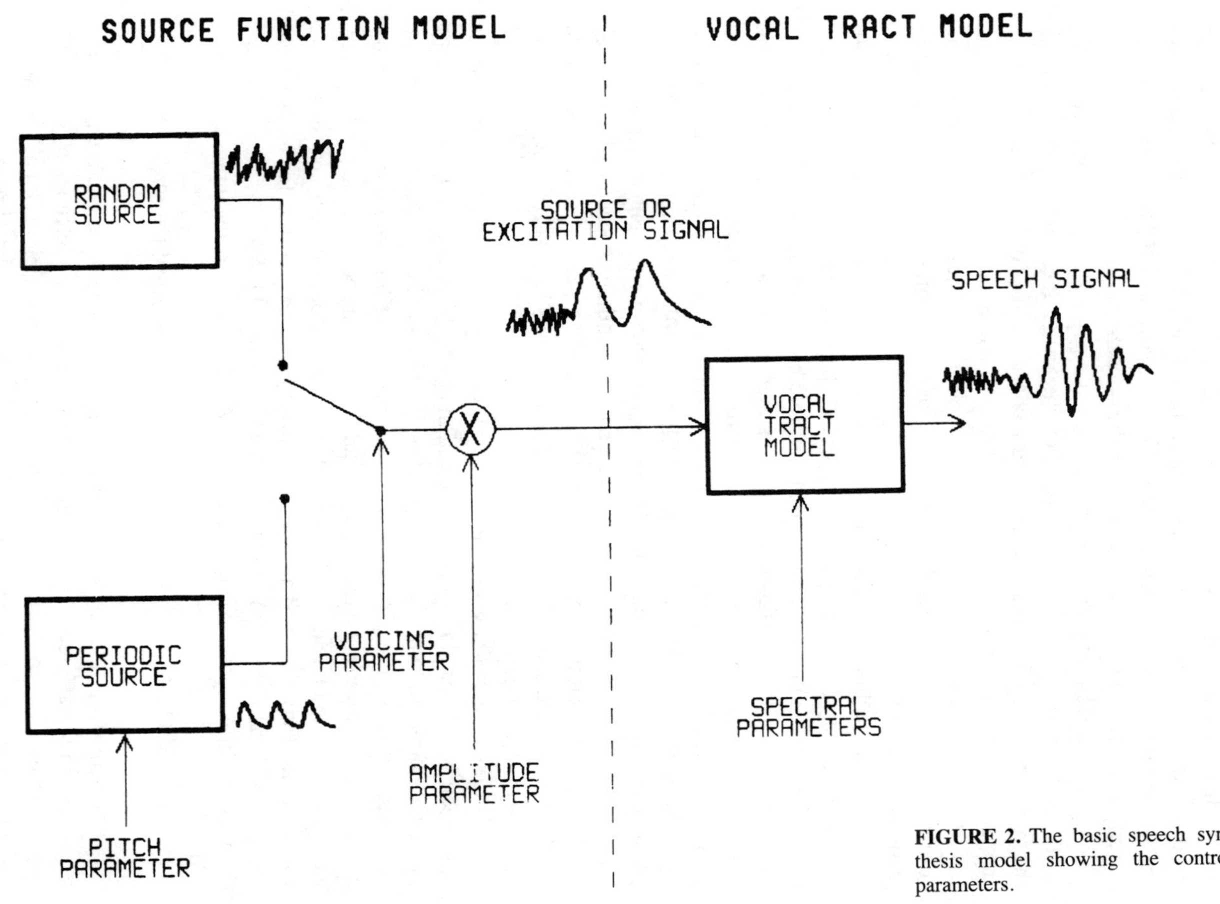

FIGURE 2. The basic speech synthesis model showing the control parameters.

the pitch parameter. In summary, the excitation signal is specified in the basic model by three parameters: an energy parameter which determines the loudness of the speech; a voiced/unvoiced parameter; and, if voiced, a pitch parameter which specifies the fundamental periodicity of the speech signal.

The second component of the model in Figure 2 is a filtering operation which imposes the proper spectral shape on the artificial excitation signal, just as the human vocal cavity controls the spectral characteristics of its excitation signal. In particular, the resonances of the vocal cavity, which result in spectral peaks in the output speech, can be accurately controlled. These spectral peaks, or formants, are a key spectral characteristic that our hearing systems use to understand speech.

Various techniques have been used to simulate the manner in which the human vocal cavity imposes a particular spectral shape on the excitation signal. One of the first techniques developed uses multiple band-pass filters. The center frequencies of the filters are fixed, but an adjustment in the gain of each filter or channel allows the desired spectrum to be approximated. In this manner, a channel synthesizer approximates the vocal tract transfer function by direct spectral measurements (see Figure 3a). The number of filters required can be reduced if it is also possible to control their center frequencies. By matching the center frequencies to the desired formant frequencies, one can generate synthetic speech with only three or four tunable bandpass filters. Since the controlling parameters in this type of synthesizer are the center frequencies of formants, this is usually called a formant synthesizer (see Figure 3b).

With the recent advances in digital processing, digital filtering techniques have been widely used to filter the excitation signal. That is exactly what is done in the technique known as linear prediction (see Figure 3c). In this approach, the synthetic speech signal is generated as the output of a filter whose input is the appropriate excitation sequence. Each digital synthetic speech sample can be generated as a weighted linear combination of previous output samples and the present value of the filter input. This yields the following expression for each output sample ($S[i]$) as a function of previous samples ($S[i-1]$, $S[i-2]$, . . . , $S[i-n]$, the prediction weights ($A[1]$, $A[2]$, . . . , $A[n]$, and the filter input ($U[i]$):

$$S[i] = A[1]S[i-1] + A[2]S[i-2] + . . . + A[n]S[i-n] + U[i]$$

The filter input is the product of the amplitude parameter and the excitation sequence. In linear predictive coding (or LPC), the filter coefficients control the spectral shape of the output signal. As the number of coefficients increases (as n is made larger in the above equation), a larger number of spectral peaks can be approximated. (For a thorough discussion of LPC, see Markel & Gray, 1976. A single-chip LPC-based speech synthesizer is described by Wiggins & Brantingham, 1978.)

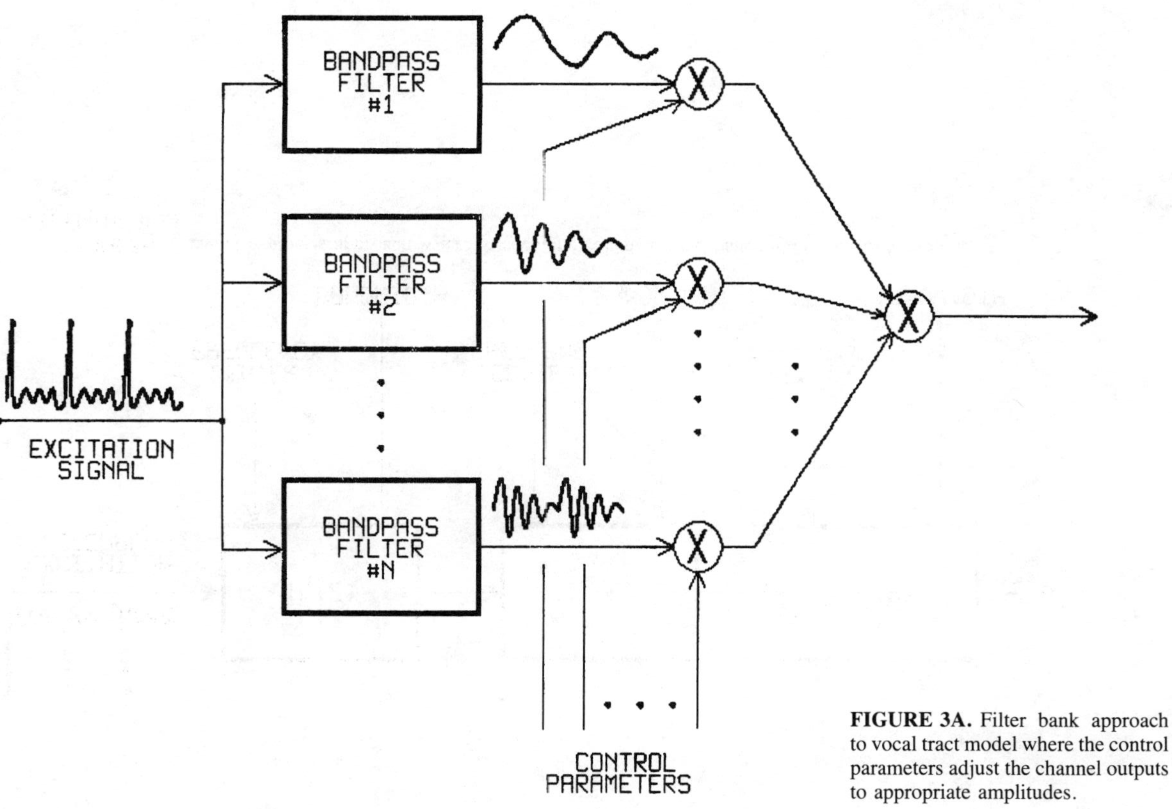

FIGURE 3A. Filter bank approach to vocal tract model where the control parameters adjust the channel outputs to appropriate amplitudes.

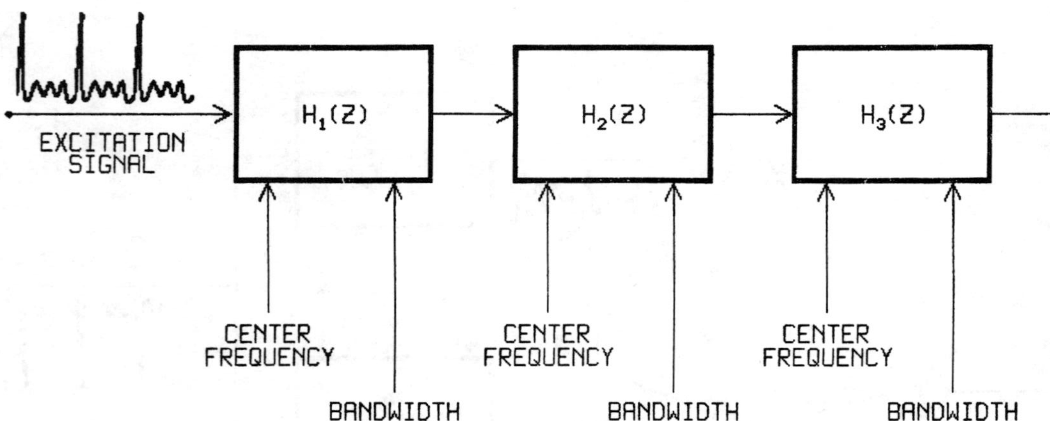

FIGURE 3B. Formant synthesis approach where the center frequencies and bandwidths control a cascade of band-pass filters.

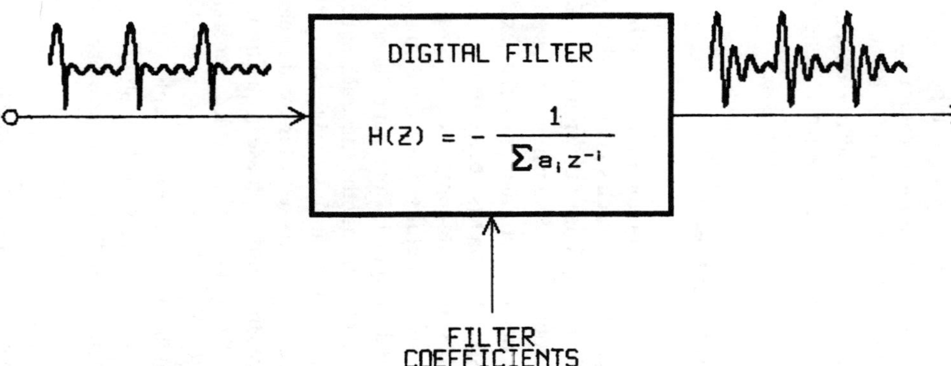

FIGURE 3C. Linear prediction uses a single higher order filter whose coefficients are chosen so that the transfer function matches that of the vocal tract.

Once the complete set of parameters (amplitude, voicing, pitch, and spectral parameters) has been specified, the speech synthesizer can produce a constant synthetic speech-like sound. Human speech, however, consists of signals with rapidly varying characteristics. It contains many short contiguous segments of voiced and unvoiced speech. Further, the spectral characteristics are constantly changing as the tongue, jaw, and lips are moved. For these reasons, the model parameters need to be updated as often as 40 to 50 times each second to generate natural sounding synthetic speech. In addition, parameter smoothing is usually employed to remove abrupt transitions. This requires that a steady stream of data be supplied to the synthesizer. These parameters can be obtained by an analysis of actual speech, or by an alternative process which is generally called synthesis-by-rule.

2.3 Analysis/Synthesis Systems

Perhaps the simplest and most commonly used method of obtaining the speech parameters is to analyze actual speech signals. In this approach, speech is recorded and a short-time spectral analysis of the signal is performed many times each second to obtain the appropriate spectral parameters as a function of time. A second analysis is then performed to determine appropriate excitation parameters. This process decides if the speech signal is voiced or unvoiced and, when it is voiced, the appropriate pitch values are computed. When the parameters controlling the synthesizer have been carefully determined, the resulting synthetic speech may sound identical to the original. Note that no attempt has been made to reproduce the original time waveform; only the spectral characteristics have been preserved. Figure 4 shows a complete analysis/synthesis system.

In the first step, a person, typically a professional speaker in an anechoic chamber, distinctly pronounces the word or phrase in question. The analog speech signal thus obtained is converted to a digital sequence, generally at a data rate of about 96K bits per second (8,000 12-bit samples per second). At this point, of course, the analog signal could be reconstructed with a digital-to-analog converter. In the second step, the digital speech analysis algorithms are used to compute the synthesizer parameters. Note that when these parameters are used in a corresponding speech synthesizer whose output is then converted into an analog signal, a different time waveform is obtained. However, the frequency content of the original speech signal is closely approximated. The speech synthesizer used could be any of several types, including a formant synthesizer or a synthesizer based on linear prediction.

The time trajectories of the speech synthesizer parameters may now be quantized in time and amplitude. Often, the parameters are periodically redefined, typically 40 or 50 times per second, by frames of data that simultaneously update all of the excitation and vocal tract parameters. Generally, coding techniques can

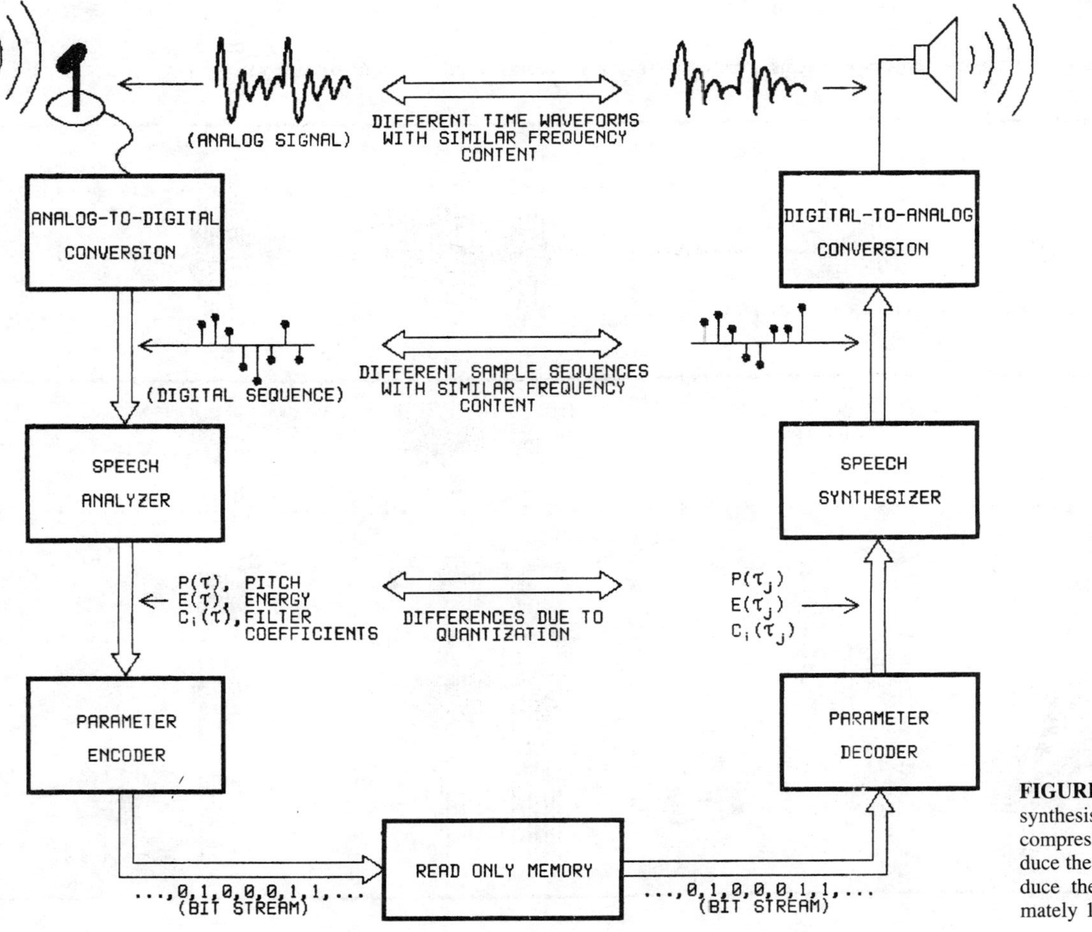

FIGURE 4. The digital analysis/synthesis system used for speech compression. These systems can reduce the data rate necessary to reproduce the speech signal to approximately 1,200 bits per second.

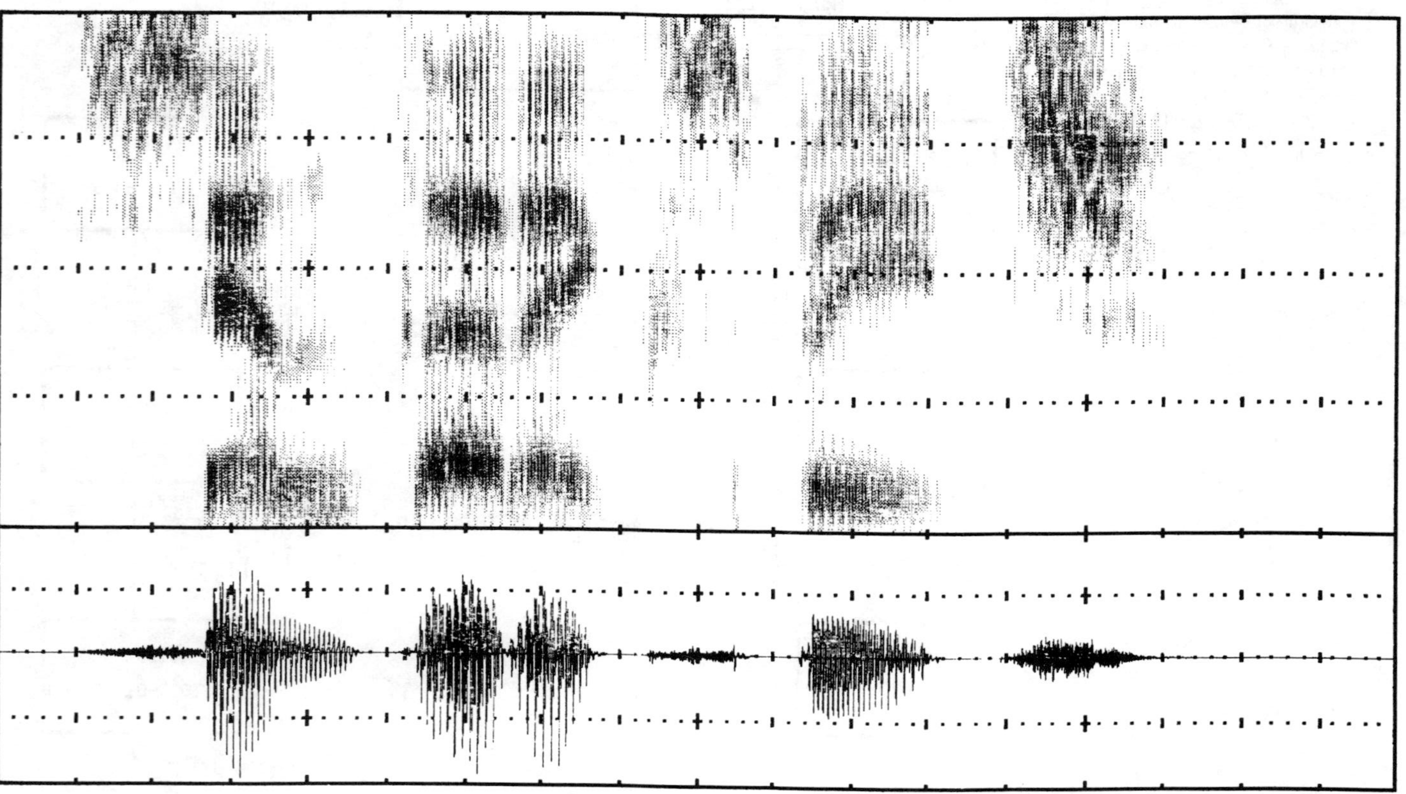

FIGURE 5A. Spectrogram and waveform of the original speech signal. (The words spoken are ''synthetic speech.'')

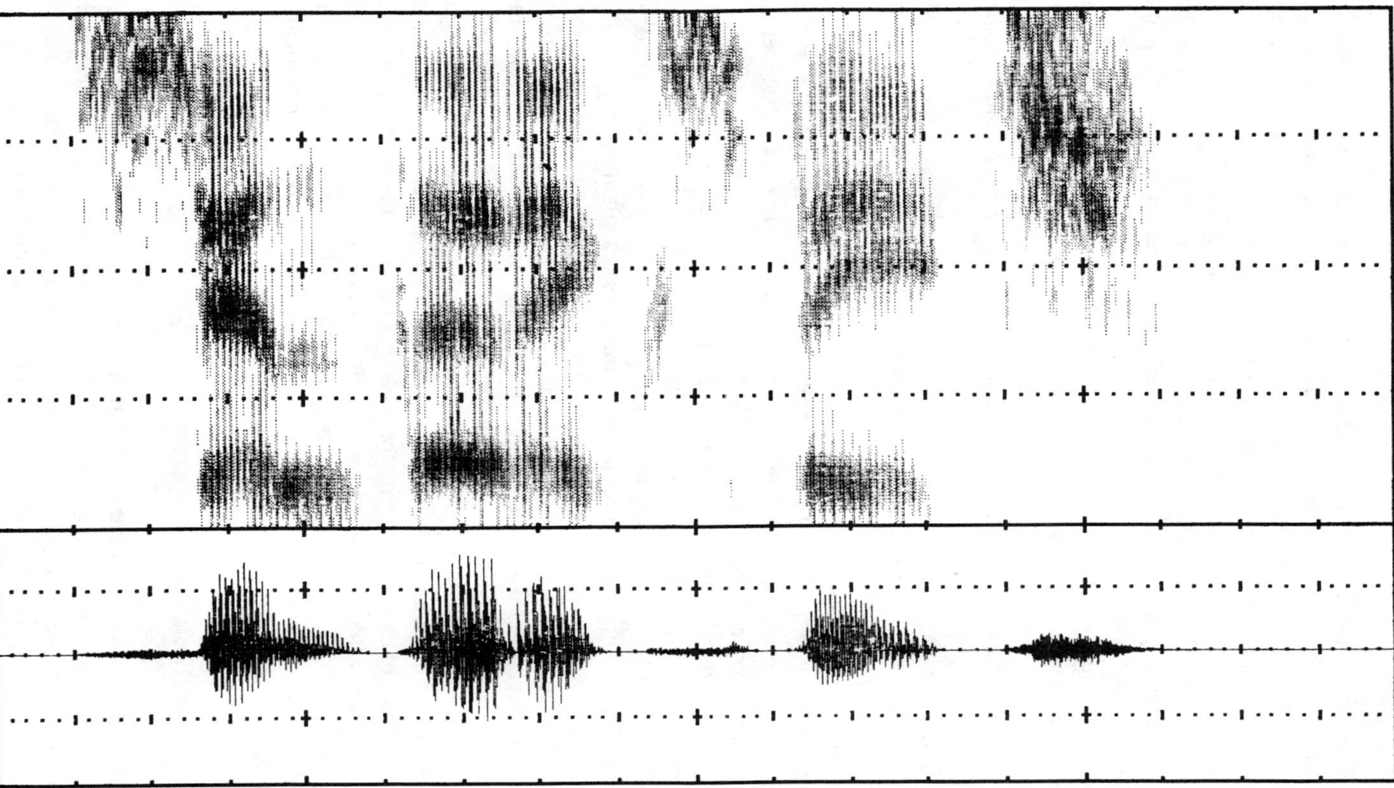

FIGURE 5B. Spectrogram and waveform of the synthetic speech signal.

reduce the bit rate to the range of 1,200 to 2,400 bits for each second of speech. The reduction in bit rate from the original digitized speech may be on the order of 100 to 1. This can represent a substantial savings in memory for voice response applications, or bandwidth for digital voice communications.

In the communications application, it is clear that the voice analysis must be done in real time. For the voice response applications, however, a great deal of time and computing resources can be used to increase the synthetic voice quality and lower the data rate. Extensive research is being performed today to improve analysis techniques and procedures for both voice communication and voice response applications. Although the synthetic voice quality and intelligibility at low data rates are not always considered satisfactory, new algorithms are constantly improving performance. Furthermore, the processing equipment required to perform this task is rapidly decreasing in size and cost.

Figures 5a and 5b compare original and synthetic waveforms for the words "Synthetic Speech," spoken by a male speaker. Above each waveform is the sound spectrogram. This is a time-frequency-intensity plot of the waveforms, and clearly shows the preservation of the speech formant structure through the analysis/synthesis procedure. In this example, a spectral analysis of the original speech was performed every 10 milliseconds, and the speech was synthesized using a twelfth order LPC model.

From a human factors standpoint, the chief disadvantage of analysis/synthesis systems is that they have a limited vocabulary. They can form messages from only those words and phrases that have been previously encoded and stored in their memories. As a result, the human factors engineer may be called upon to reduce a system's storage costs by determining the minimum user-acceptable size and content of the system's vocabulary. (This problem is addressed in Section 3.3 of this chapter.) Further, unless context is carefully controlled, messages composed of concatenated words and phrases can sound very awkward and unnatural, even though intelligibility may be excellent. In this case, the general pitch, time, and cadence of the speech—its prosody—will be incorrect. One solution is to modify the prosody of the concatenated segments according to their context. This, of course, requires modification of the acoustical parameters that form the input to the speech synthesizer. Now, instead of reading these parameters directly from memory, the synthesizer must define some or all of the parameters at the time of synthesis.

2.4 Synthesis-By-Rule Systems

A synthesis-by-rule system is illustrated in Figure 6. In these systems, the parameter composition module determines the acoustic model parameters at the time of synthesis. Usually, more than just the prosody is considered. To achieve natural-sounding speech, it is necessary to consider the problem of coarticulation.

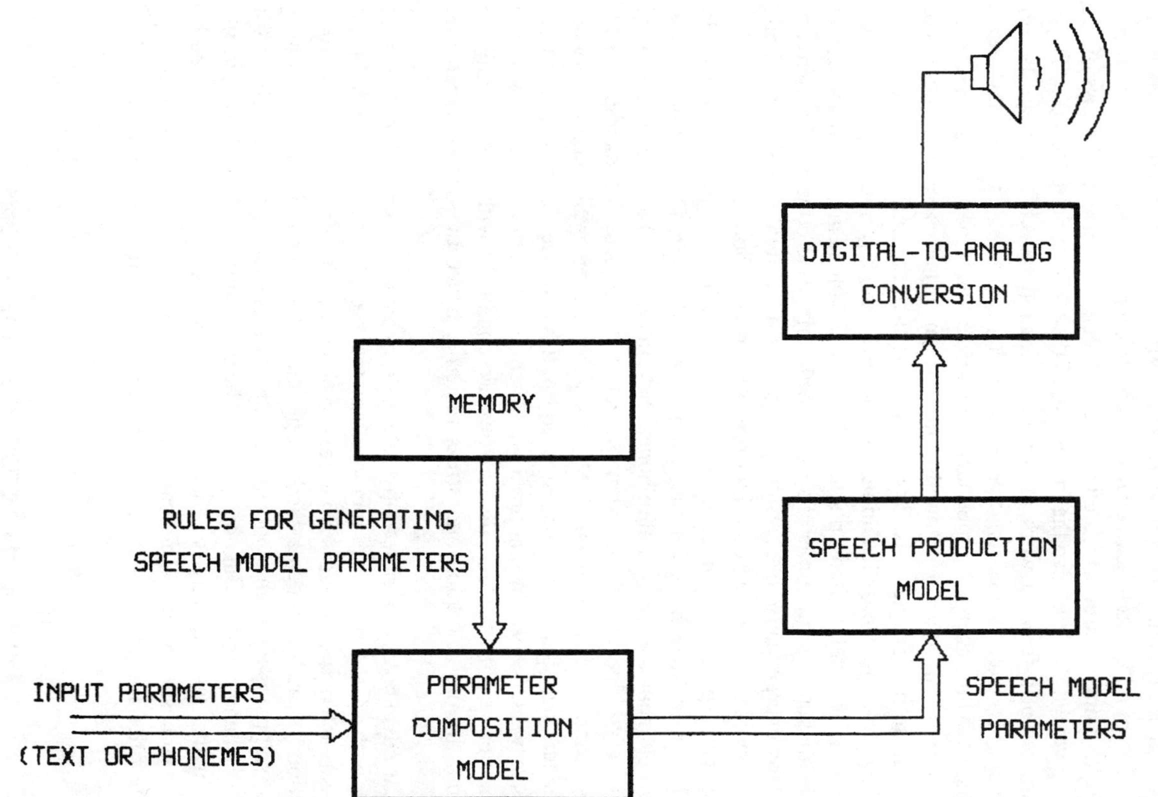

FIGURE 6. A synthesis-by-rule system generates speech model control parameters by applying a set of rules to some input parameters, such as text or phonemes.

Coarticulation is a modification of pronunciation due to the influence of neighboring sounds. For example, the word ''your'' in ''have it your way'' sometimes resembles ''chore,'' as in ''have it chore way.'' In the phrase ''sit down'' the unvoiced ''t'' sound is often lost because of the influence of the following voiced ''d.'' In the phrase ''used to'' the unvoiced ''t'' causes not only the loss of the preceding voiced ''d'' but also the unvoicing of the ''z'' sound in ''used.'' These are examples of a force at work in normal colloquial speech that linguists call ''economy of effort''—that is, the tendency to say things the easiest way. Whereas the synthesizer will pronounce ''bookcase'' as ''book'' plus ''case,'' in natural speech only one ''k'' sound is uttered. In instances such as these, the synthesizer, to its detriment, is far more precise than the human tongue.

This problem of coarticulation becomes even more serious when one considers what is generally called ''phoneme'' synthesis. Phonemes are considered to be the smallest perceptual units of speech, with only 40 to 60 in the English language. However, because of coarticulation problems, natural-sounding synthetic speech cannot be generated by straightforward concatenation of only 40 to 60 different speech segments. In order to sound natural, such synthesizers must modify every elementary speech segment according to a set of coarticulation rules. These modified phonemes are called allophones. Several systems have been developed that synthesize speech by concatenating allophones (see, for example, Lin, Goudie, Frantz, & Brantingham, 1980). Systems of even greater complexity are able to synthesize speech from unrestricted English text (see, for example, Allen, 1976; Kurzweil, 1979). In this process, which utilizes extensive linguistic and syntactic analysis of the text, dictionaries are combined with a set of rules to determine the pronunciation and stress patterns, hence the descriptive name, synthesis-by-rule.

The advantages of synthesis-by-rule systems are that larger vocabularies are possible with a fixed amount of memory, and that speech may be generated from more easily interpreted inputs such as text or phonemes. This means that, in some applications, ''raw'' input data can be supplied by the user at the time of synthesis. An example is a portable phonetic synthesizer that serves as a speech aid for the vocally handicapped (Gagnon, 1978). The disadvantage of synthesis-by-rule systems is that voice quality is generally not nearly as good as that produced by analysis/synthesis systems. The reason is that it is extremely difficult to find appropriate rules for pitch, timing, and stress that will produce natural-sounding speech.

3. HUMAN FACTORS CONSIDERATIONS

Two major human factors considerations in the use of speech synthesizers are voice quality and vocabulary limitations. Sections 3.2 and 3.3 of this chapter address those issues. The fundamental human factors consideration, however, centers on when to use speech synthesizers.

3.1 When Should Speech Synthesizers Be Used?

Nearly all general textbooks of human factors engineering discuss at length when to use auditory or visual forms of presentation (see, for example, Chapanis, 1965, chaps. 3 and 4; McCormick, 1976, part 2; Woodson & Connover, 1964, chap. 2). Although many of the guidelines presented in these references were originally developed over 30 years ago, they still apply to modern human-machine interfaces. For convenience, these guidelines are summarized below.

In general, a visual output mode, such as a CRT or a hard-copy printer, is preferable when:

1. The message is complex, contains technical or scientific terms, or uses terms with which the user might not be familiar.
2. The message is long.
3. The message needs to be referred to later. (In this case, the hard-copy printer would be better than the CRT, which would have no real advantage over speech output.)
4. The message deals with spatial orientation or the location of points in space. (Such information is usually best presented graphically. For this sort of application, an alphanumeric display would have no advantage over speech output.)
5. There is no urgency in the message.
6. The auditory channels are overloaded with messages, signals, or sounds to which the user must pay attention.
7. The auditory environment is too noisy or otherwise unsuitable for the reception of aural messages.
8. The user may remain in a position where he or she can easily see the displayed messages.
9. The system output consists of many different kinds of information which must be displayed simultaneously and which must be monitored and acted upon by the user.

On the other hand, speech output is frequently preferable when:

1. The message is simple.
2. The message is short.
3. The message will not be referred to later.
4. The message deals with events in time.
5. The message requires an immediate response.
6. The visual channels of communication are overloaded.
7. The environment may be too brightly lit, too poorly lit (possibly to preserve dark adaptation), subject to severe vibration, or otherwise unsuitable for transmission of visual information.
8. The user must be free to move around.

9. The user may be subjected to high G forces or anoxia (lack of oxygen, typically at high altitudes). The magnitude of the G forces or anoxia at which eyesight begins to be impaired is well below that needed to affect hearing.

In reviewing these guidelines, the human factors engineer should recognize that they are incomplete, in that they are oriented exclusively toward environmental or task considerations. In evaluating the merits of a particular application for speech synthesizers, or when developing a new application, human needs must also be considered. Chief among these considerations is that people are different, and certain user populations will have special needs. Indeed, most of the successful synthesizer-based products presently on the market were designed to meet the special needs of a particular set of users, rather than the requirements of a particular job or environment. For example, speech synthesizers are being used in products for people who cannot read printed material (see, for example, Kurzweil, 1979; Songco, Allen, Plexico, & Morford, 1980). They can also be the voices of the vocally handicapped (see, for example, Gagnon, 1978). They have even found their way into electronic hand-held learning aids for children, probably their most widely used application to date. Examples of these children's learning aids include the Texas Instruments Speak & Spell, Speak & Read, and Speak & Math learning aids, which are intended to teach specific skills through drill and practice. ("Speak & Spell," "Speak & Read," and "Speak & Math" are registered trademarks of Texas Instruments Incorporated. The Speak & Spell was the first product in this series; see Frantz & Wiggins, 1981, for a discussion of the human factors that were considered in its development.)

The point is that this is not the sort of technology that can be fully exploited simply through the application of "cookbook" human factors guidelines. While such guidelines are a very useful starting point, the key ingredient in this recipe is a knowledge of people's needs. In short, speech synthesizers should not be thought of solely as a way to enhance existing interfaces; entirely new classes of interfaces and products are possible, limited only by the skill and creativity of the designer.

Once a good application for speech synthesis technology has been identified, the human factors engineer will then be concerned not only with the classical problems associated with any speech-output interface, but also with a new set of problems unique to this technology. The bulk of these problems relate to voice quality and vocabulary selection.

3.2 Voice Quality

One of the goals of human factors engineers who work with speech communication systems is to decrease the likelihood that listeners will misunderstand or confuse words. The previously cited human factors textbooks discuss factors in communication systems and listening environments that can contribute to such errors, including a poor overall signal-to-noise ratio, acoustic masking by certain narrowband noises, echoes and reverberations, and both spectral and amplitude distortions of the signal. These texts also recommend solutions for listening problems attributable to these factors. While the recommended procedures are certainly applicable to portions of synthesizer-based systems, the synthesizers themselves can distort speech in ways that cannot be corrected by these techniques. This requires human factors engineers to apply new techniques, as well as old, to synthesizer-based systems.

Probably the best review of the voice-quality problems of synthesizers is by Voiers (1977), who performed factor analyses of people's ratings of synthetic voices, and identified the independent factors that contributed to their perception of the overall voice quality. Although Voiers' subjects evaluated different real-time analysis/synthesis telecommunication systems, many of his findings also apply to analysis/synthesis systems that synthesize speech from data stored in their memories. He reported that, for the analysis/synthesis systems he tested, actual intelligibility accounted for only about two thirds of the variance in his subjects' ratings. The remaining one third of the variance was attributed to perceived intelligibility and aesthetic qualities.

With regard to aesthetics, it has also been the authors' experience that seemingly intangible qualities have a large impact on people's evaluation of synthetic voices. In one case, a particular voice was selected by TI engineers because it sounded pleasant and helpful. This same voice was then rejected by the majority of test subjects, many of whom described it as sounding evil and sinister. It is clear that, for the time being, many people have an emotional reaction to synthesized speech. At least until the public becomes more familiar and comfortable with synthetic speech technology, human factors engineers need to be sensitive to these concerns.

The remainder of this section describes ways to improve the intelligibility of present-generation synthesizers solely by modifying stored speech production parameters. Recall that there are four sets of speech production parameters: the voiced/unvoiced decision, amplitude, pitch (if voiced), and the spectral components. For all practical purposes, any usual speech sound can be defined in terms of these parameters. Therefore, if the output of a well-designed synthesizer is not intelligible, it is because the synthesizer was not fed the proper parameters for the desired sound.

In describing ways to improve a synthesizer's intelligibility, this section concentrates on techniques the authors have used to improve the intelligibility of LPC-based analysis/synthesis systems. However, these techniques can be applied to many other synthesis methods. Most of the techniques require trained people to edit and make corrections to speech production parameters manually before they are stored in a product's memory. It cannot be overstressed that the most accurate and rapid editing is accomplished using a fully interactive system that displays the parameters frame-by-frame in a simple table on a CRT. This allows the editor to easily modify any parameter, and to run both the original and modified parameters through a synthesizer to compare how they sound.

It has been the authors' experience that errors in the amplitude, voiced/unvoiced, and pitch parameters are the most common source of voice-quality problems. Fortunately, because an error in one of these parameters creates its own unique type of audio distortion, it is relatively easy to identify and correct the faulty parameter. Errors in the spectral parameters are another matter. Even for highly skilled editors, the identification and correction of faulty spectral parameters can be an extremely tedious task, largely because it is difficult to "visualize" how changing a particular spectral parameter will affect the resulting sound.

While the sort of parameter errors mentioned above are common to all speech sounds, plosives also have a set of unique problems. These are due to the fact that the synthesizer's modeling of the human vocal tract as a function of time is discrete, rather than continuous. For example, in the speech synthesizer used with the Texas Instruments 99/4 computer, the speech production parameters are updated and changed every 25 milliseconds. It may be convenient for the reader to consider this process as the acoustic equivalent of a motion picture projected at 40 frames per second, except, of course, that there are no pauses equivalent to those in motion pictures when the film is being advanced to the next frame. The problem this creates for synthesizers is that, as is the case with motion pictures, events that occur in much less time than the frame length become blurred.

Plosive sounds in human speech typically have a duration of about five milliseconds. To the human ear, it is a sudden, brief jump in amplitude that gives these sounds their plosive quality, while their spectral characteristics allow the ear to distinguish one plosive from another. If a plosive sound's spectral characteristics were synthesized for a longer than usual duration, and if there were no characteristic brief jump in amplitude, the resulting sound could easily be misinterpreted by listeners as a fricative. An example of such a problem was found in the initial version of the Texas Instruments Speak & Spell, where informal field testing of the product found that some people thought the B in "obey" sounded more like a V.

The solution to this problem requires changing amplitude parameters in order to give the sound more of a plosive character. Although the duration of the sound is limited by the synthesizer's frame rate, and therefore cannot be shortened to its natural duration in human speech, this does not preclude synthesis of sounds that are psycho-acoustically equivalent to plosives. The sudden amplitude changes characteristic of plosives can be duplicated by synthesizers and, despite the longer-than-normal duration of the sound, these amplitude changes are sufficient to give the sound a plosive quality. To accomplish this, the speech production data for a particular word or allophone must be examined to determine which frame contains what is supposed to be a plosive. Once that frame is identified, its amplitude parameter should be adjusted upward, while the amplitude parameters of the adjacent frames should be adjusted downward. The amount of adjustment necessary will vary depending on many other factors, but this technique will usually restore the plosive quality to these sounds.

The difficulties with plosives do not end here, however. In analysis/synthesis systems, the speech production parameters are computed from a digital recording of a person saying the needed word. In order to obtain the speech production parameters, the digital recording is first partitioned into frames (or short segments of speech), and then the speech model parameters are computed for each frame. This process computes the parameters averaged across the entire frame, and events that comprise a small portion of the frame, such as plosives, can get lost in the "noise" as their spectral characteristics are averaged into those of the other sounds in that frame. If this happens, even the amplitude adjustments described in the previous paragraph will not by themselves restore the proper sound. There are two solutions to this problem. The first requires the editor to correct the spectral parameters manually, a tedious job that is very difficult to do accurately. The second solution is much easier: in the initial recording of the word, if the speaker exaggerates the plosive sounds without changing the prosody of the word, just as a skilled stage actor might do, this may make it less likely that the plosive's spectral characteristics will be lost in the noise. It is a relatively simple matter later on to adjust the amplitude parameters in order to keep the plosives from sounding unnaturally exaggerated in the final product.

A similar problem can occur when, in dividing the digital speech recording into frames prior to analysis, the system puts a frame border in the middle of a plosive. The result is that the spectral characteristics of the plosive now must overcome the "noise" of two frames rather than one. The solutions are the same as those recommended in the previous paragraph, except that there is an additional step that may be tried: the digital recording may be reanalyzed with the frame borders shifted slightly so that all plosives fall entirely within single frames.

3.3 Vocabulary Limitations

As was previously stated, the major human factors drawback of analysis/synthesis systems is that they cannot synthesize words not previously stored in their memories. By contrast, synthesis-by-rule systems that synthesize speech from English text (rather than user-generated phonemic or allophonic input) are designed to "sound out" words based on their spelling by applying letter-to-sound rules. Unfortunately, many English words are not pronounced the way they are spelled. If good voice quality is a requirement, synthesis-by-rule systems usually require a "dictionary" in their memories in which to look up the proper pronunciation of these words. So, as is the case with analysis/synthesis systems, synthesis-by-rule systems with acceptable voice quality must also be limited by the number of rules and dictionaries their memories can hold.

Although human factors engineers may desire a large vocabulary, adding words to a synthesizer's memory also adds to its expense. This is especially true of anaylsis/synthesis systems, where the incremental memory space requirements for a given word are much greater than they are in synthesis-by-rule systems. This is because analysis/synthesis systems store words as the parameters that control the speech production model, whereas the "dictionaries" of synthesis-by-rule systems merely contain pronunciation rules from which the speech production parameters are computed. The result, in essence, is a new version of an old problem for human factors engineers: How much can the interface's price and bandwidth, in this case its vocabulary, be reduced without adversely affecting performance? When designing a synthesizer-based interface, this question becomes: What is the minimum user-acceptable vocabulary size? What should those words be?

This issue was, of course, first explored in the 1930s by linguists who developed highly simplified, general purpose subsets of English, such as the 1700-word vocabulary Basic English (Ogden, 1933; Johnsen, 1944). The Basic English vocabulary was later successfully used as the lexicon for SAD SAM—one of the first computer programs to accept English language data entry and answer English language questions about relationships in the data (Lindsay, 1963). However, Basic English was intended by its designers to be a complete, stand-alone language that could be used in a wide variety of situations. How large must a vocabulary be if it is intended solely for a particular type of situation or application? Surprisingly small: a laboratory study of interactive limited vocabulary dialogue by Kelly and Chapanis (1977) has demonstrated that carefully chosen vocabularies as small as 300 words can be just as efficient as unrestricted vocabularies for many kinds of applications. An examination of the Kelly and Chapanis protocols by Michaelis, Chapanis, Weeks, and Kelly (1977) suggests that even smaller vocabularies could have been used successfully. This is not meant to imply any restricted subset of English will work; it cannot be overstressed that high efficiency and user acceptability can be expected only with a carefully chosen vocabulary.

With regard to the actual selection of the words to be included in a synthesizer's vocabulary, one of the major contributions of the Kelly and Chapanis study was the idea of having separate vocabulary lists, one a general vocabulary resident in the system that would be used for all applications, and the other an application-specific vocabulary that would be changed, depending on the application. This is the design later adopted by Texas Instruments Incorporated for their 99/4 home computer speech synthesizer. This synthesizer has a resident general vocabulary of about 365 words stored in about 340 separate memory locations. The number of words is greater than the number of memory locations because homophones, such as to, too, and two, have the same speech model parameters, and therefore share memory locations. In addition to words describing some aspect of the computer, such as cassette, console, data, and diskette, words were selected for inclusion in this synthesizer's resident vocabulary based on their frequency of use by the restricted subjects in the Kelly and Chapanis study, as reported in the Michaelis et al. (1977) reference. These words are not intended as a stand-alone vocabulary; they need to be supplemented by plug-in modules containing application-specific vocabularies. An example of a currently available application-specific vocabulary module for the TI 99/4 is a module that supplements the general vocabulary with such things as the names of some animals and nursery-rhyme characters, thereby allowing the computer to tell children's stories.

TI has years of experience preparing software using the 99/4 speech synthesizer's general vocabulary. A variety of programs have been written, and records have been kept on the frequency of use of words from that vocabulary, just as records have been kept of cases where programmers needed words that were not available. From these records, a list of words has been compiled that the authors recommend be included in the resident vocabularies of most synthesizers. The word list is shown in Table 2. The length of this list is such that, with current techniques, all of the words can be stored in LPC format on a single 16-kilobyte memory chip which can be inexpensively mass produced and used in synthesizers over a wide range of applications. It is important to note that this list only roughly corresponds to lists of the most commonly used words in unlimited dialogue (see, for example, Kucera & Francis, 1967).

The human factors engineer must still, of course, be concerned with the composition of application-specific vocabularies to be used with this general vocabulary. Once again, the basic requirement is to develop a list of all words that are essential for communication, while minimizing the storage space requirements in the system's memory. This can be a difficult task if the system designers have only a vague idea of how the interface will operate, because some unnecessary words will probably be included, while some needed words will be left out.

It is the authors' experience that the best first step in determining the content of an application-specific vocabulary is to prepare scripts of what an unlimited synthesizer's output would be under all of the application's anticipated scenarios. In some cases, the designers of the interface are able to prepare good scripts based on their knowledge of how the interface and system will be used. Frequently, though, the authors have found it necessary to obtain scripts by recording simulations of the interface during all of the anticipated scenarios. The simulations are accomplished by having one or more people play the role of the speech synthesizer. Although no attempt is made to limit their choice of words, these people are generally instructed to keep their verbal output as concise as possible; this tends to greatly reduce their use of unnecessary words without detracting from their ability to communicate (see Ford, Weeks, & Chapanis, 1980; Michaelis, 1980). It is, of course, preferable to run the simulations with a number of different people in the roles of synthesizer and operator so that the scripts will not be biased toward any one person's vocabulary or style.

Once the scripts are obtained, a list should be made of all the different words that were used, rank ordered by their frequency of use. Eliminate words already

The Recommended Core Vocabulary for Speech Synthesizers

A(letter A)	FIFTY	LOOKS	S	TIMES
A (''uh'')	FIND	M	SAME	TO/TOO/TWO
ABOUT	FIRST	MAKE	SAY	TOO/TO/TWO
ALL	FIVE	ME	SAYS	TRY
AM	FOR/FOUR	MEAN	SEE/C	TURN
AN	FORTY	MESSAGES	SET	TWELVE
AND	FOUR/FOR	MILLION	SEVEN	TWENTY
ANSWER	FOURTEEN	MORE	SEVENTEEN	TWO/TO/TOO
AS	FROM	MUST	SEVENTY	U/YOU
AT	G	MY	SHOULD	UNDERSTAND
B/BE	GET	N	SIX	UNTIL
BAD	GIVE	NEGATIVE	SIXTEEN	UP
BE/BE	GO	NEXT	SIXTY	USE (''uze'')
BETWEEN	GOOD	NINE	SMALL	V
BLACK	GREEN	NINETEEN	SO	VERY
BLUE	H	NINETY	SOME	W
BOTH	HAS	NO/KNOW	SORRY	WAIT
BUT	HAVE	NOT	START	WANT
BUTTON	HEAR/HERE	NOW	STEP	WAS
BUY/BY	HELLO	O	SUPPOSED-TO	WE
BY/BUY	HELP	OF	SURE	WELL
C/SEE	HERE/HEAR	OFF	T	WERE
CAN	HUNDRED	OKAY	TAKE	WHAT
COME	I	ON	TELL	WHERE
CORRECT	IF	ONE	TEN	WHICH
D	IN	ONLY	THAN	WHITE
DECIMAL	IS	OR	THAT	WHY/Y
DID	IT	OTHER	THE (''thee'')	WILL
DIFFERENT	IT'S/ITS	P	THE (''thuh'')	WITH
DO	ITS/IT'S	POINT	THEIR/THERE	WORK
DOES	J	POSITION	THEM	WOULD
DONE	JUST	POSITIVE	THEN	WRITE/RIGHT
DOWN	K	POWER	THERE/THEIR	WRONG
E	KEY	PRESS	THEY	X
EIGHT	KNOW/NO	PROBLEM	THING	Y/WHY
EIGHTEEN	L	PUT	THINGS	YELLOW
EIGHTY	LARGE	Q	THIRTEEN	YES
ELEVEN	LEFT	QUESTION	THIRTY	YET
EMERGENCY	LESS	R	THIS	YOU/U
END	LET	READ (''reed'')	THOUSAND	YOUR
EXACTLY	LETS	READY	THREE	Z
F	LIKE	RED	THROUGH	ZERO
FIFTEEN	LOOK	RIGHT/WRITE	TIME	

appearing in the general vocabulary list, and the words that remain constitute an application-specific vocabulary for the application that had been simulated. However, this list often contains many more words than are necessary, and it is the job of the human factors engineer to eliminate the unnecessary words. Unfortunately, because each application has its own unique requirements, there are no hard and fast rules or cookbook approaches to follow. In general, though, the authors have found that the bulk of the words that can be eliminated are not words that had nothing to do with the application; rather, they are words that are redundant because they can be defined using one or more of the other words on the list.

When a word can be defined using just one other word, the two words are synonyms. The word lists obtained from scripts often contain many sets of what are apparently synonyms. However, before actually classifying a set of words as synonyms, the scripts must be examined to confirm that the words were used in a completely interchangeable manner. Once that is done, work can begin on eliminating all but one of the words from each synonym set. In some cases, fairly simple guidelines can be followed. If the synonyms in a set have significantly different frequencies of use, it is almost always best to keep the one that is most commonly used. On the other hand, when synonyms have similar frequencies of use, it is generally preferable to keep the shortest one because it requires less memory space. Many times, though, it is not possible to rely on guidelines, and judgments must be based on a knowledge of the system's needs. This often involves making trade-offs between the needs of the operator and the cost of the system. For instance, particularly when trying to maintain operator vigilance during a repetitive task, it would be a good idea to retain some synonyms in order to avoid monotony in the wording of messages, even though this adds to the system's memory size and cost. Sometimes it is even necessary to make what are, in essence, linguistic trade-offs. For example, some words that are synonyms may also be homonyms or homophones of other words; depending on the circumstances of their use, they may be preferable because of their added versatility, or they may be less desirable because their use might cause confusion.

A similar problem is faced by the human factors engineer when he or she must decide whether to eliminate from the list individual words that can be defined using two or more other words that are on the list. An example of such a situation can be found in a vocabulary that might be prepared for use in an aircraft cockpit: given that the words ''altitude'' and ''indicator'' are already in the vocabulary, can the word ''altimeter'' be deleted in order to save memory space? The authors approach such questions by considering the single word and the phrase that replaces it to be synonyms, in this case ''altimeter'' and ''altitude indicator.'' In all likelihood, the scripts would show that ''altimeter'' was used far more often than the phrase ''altitude indicator.'' Although most pilots would certainly understand what is meant by ''altitude indicator,'' if this phrase was not commonly used in the scripts, it would be reasonable to assume that the use of this phrase in place of ''altimeter'' might create an error-provocative situation, especially in times of stress. So, as is the case with single-word synonyms that have very different frequencies of use, it is best to keep the synonym that has been used most often.

After words have been pared from the original word list, the list must be tested to make sure that it still contains all of the necessary words. This generally requires running simulations again, but this time the people who are playing the role of the synthesizer do not have the freedom to choose their own words; rather, they must compose their messages using only words from the list, just as the subjects in the Kelly and Chapanis (1977) study did. The human factors engineer should carefully observe the simulations, looking for incidents where communication became awkward or even impossible because certain words were not available. If this happens, the lists need to be modified and the simulations repeated. It will usually take just one or two iterations to obtain a vocabulary list for a specific application that meets the goal of minimizing memory storage requirements without sacrificing the performance of the man-machine system.

3.4 Special Effects

If used with imagination, the present generation of electronic speech synthesizers can be remarkably versatile tools. Although designed to synthesize human-like speech, they are also capable of synthesizing many other sounds, such as the musical tones used in the Texas Instruments Speak & Spell. These tones, and other sounds, such as gunshots or the ''chugging'' of a steam locomotive, can be synthesized by an unmodified synthesizer if it is given the appropriate control parameters.

One useful special sound effect cannot be obtained in most synthesizers until minor modifications are made to their circuitry. Synthesizers can produce speech that is highly intelligible, yet not at all human-like. A synthesizer's normal human-like sound is largely due to its use of a periodic function generator that models the human-voiced excitation signal. However, other periodic sources may successfully be used in place of the human-like source. If these sources are properly selected, the synthetic voices thus generated sound very unusual, although they remain highly intelligible. This is not done as a gimmick: a different excitation signal may be assigned to special messages, such as warning messages, in order to improve their attention-getting capability.

4. CONCLUSION

This chapter describes the *current* generation of electronic speech synthesizers and addresses some issues relevant to their use in human-machine systems. In closing, it is worthwhile to take a brief look at some of the trends in speech synthesis technology.

In analysis/synthesis systems, the voice quality at equivalent data rates is being improved. Concurrently, the data rate necessary for good quality speech is being decreased, meaning that more words can be stored in the same amount of memory. In addition, if the prices of semiconductor memories continue to drop, the cost-per-word in all types of synthesizers will decrease even further.

The voice quality of synthesis-by-rule systems is benefiting from the refinement and expansion of the rules and algorithms that determine pronunciation. Work is even proceeding on synthesizers that will automatically add appropriate intonation contours to their messages (Pierrehumbert, 1981).

Many of these improvements in synthesis techniques do not involve a modification of the basic speech production model described in this chapter. One modification to the encoding of the control parameters, however, has yielded significant improvements in voice quality at low data rates. As is described in Section 3.2, a factor that tends to limit the voice quality of the current generation of speech synthesizers is their fixed frame rate, which is typically set at 40 to 50 frames per second. Because the duration of each set of parameters is 20 to 25 milliseconds, these synthesizers cannot faithfully reproduce the plosive sounds of human speech, which have a very short duration of about five milliseconds. One solution is a synthesizer that has a variable frame rate. In such a synthesizer, in addition to the amplitude, voicing, pitch, and spectral parameters, each frame of speech production data contains a parameter defining its duration. Chen (1982) describes an analysis/synthesis system that automatically adjusts frame durations to minimize the number of frames needed for high quality reproduction. The quality of synthesized plosive sounds is improved because they have shorter durations. Further, only one frame is needed to describe the long-duration speech sounds that typically require several frames in fixed-rate synthesizers. The result is that this synthesizer requires an average of 20 frames per second for high quality speech, half the minimum frame rate for equivalent quality fixed-rate synthesis.

In summary, advances in speech synthesis techniques are improving voice quality while reducing costs. The result is that speech synthesizers will play an increasingly important role in human-computer interaction.

5. ACKNOWLEDGEMENTS

The authors wish to thank Professor Alphonse Chapanis, Director of The Johns Hopkins University Communications Research Laboratory, and Ramona R. Michaelis, Supervising Editor of the Funk & Wagnalls Standard College Dictionary, for their helpful reviews and revisions of this manuscript.

6. REFERENCES

Allen, J. Synthesis of speech from unrestricted text. *Proceedings of the IEEE*, 1976, *64* (4), 433–442.
Chapanis, A. *Man-machine engineering*. Belmont, CA: Brooks/Cole, 1965.

Chen, Y. An analysis-by-synthesis approach for automatic time segmentation of speech signals. *Proceedings of the IEEE International Conference on Acoustics, Speech, and Signal Processing*, Paris, France, May 1982.

Ford, W. R., Weeks, G. D., & Chapanis, A. The effect of self-imposed brevity on the structure of dyadic communication. *The Journal of Psychology*, 1980, *104*, 87–103.

Frantz, G., & Wiggins, R. H. The development of "solid state speech" technology at Texas Instruments. *The Newsletter of the IEEE Acoustics, Speech, and Signal Processing Society*, March 1981, *53*.

Gagnon, R. T. Votrax real-time hardware for phoneme synthesis of speech. *Proceedings of the IEEE International Conference on Acoustics, Speech, and Signal Processing*, Tulsa, OK, April 1978.

Johnsen, J. E. (Ed). *Basic English*. New York: The W. H. Wilson Company, 1944.

Kelly, M. J., & Chapanis, A. Limited vocabulary natural language dialogue. *International Journal of Man-Machine Studies*, 1977, *9*, 479–501.

Kucera, H., & Francis, W. N. *Computational analysis of present-day American English*. Providence, RI: Brown University Press, 1967.

Kurzweil, R. Kurzweil talking terminal announced; first unlimited vocabulary. *The Kurzweil Report*. Kurzweil Computer Products, Cambridge, MA. Spring 1979.

Ladefoged, P. *A course in phonetics*. New York: Harcourt Brace Jovanovich, 1975.

Lin, K., Goudie, G., Frantz, G., & Brantingham, L. Text to speech using LPC allophone stringing. *Proceedings of the Fall IEEE Consumer Electronics Conference*, Chicago, November 1980.

Lindsay, R. K., Inferential memory as the basis of machines which understand natural language. In E. A. Feigenbaum & J. Feldman (Eds.), *Computers and thought*. New York: McGraw-Hill, 1963.

Markel, J. D., & Gray, A. H. *Linear prediction of speech*. New York: Springer-Verlag, 1976.

McCormick, E. J. *Human factors in engineering and design* (4th ed.). New York: McGraw-Hill, 1976.

Michaelis, P. R. Cooperative problem solving by like- and mixed-sex teams in a teletypewriter mode with unlimited, self-limited, introduced, and anonymous conditions. *JSAS Catalog of Selected Documents in Psychology*, 1980, *10*, 35–36. (Ms. No. 2066)

Michaelis, P. R., Chapanis, A., Weeks, G. D., & Kelly, M. J. Word usage in interactive dialog with restricted and unrestricted vocabularies. *IEEE Transactions on Professional Communication*, 1977, *PC-20*, 214–221.

Ogden, C. K. *Basic English: an introduction with rules and grammar* (4th ed.). London: Kegan Paul, Trench, Trubner & Co., 1933.

Pierrehumbert, J. Synthesizing intonation. *Journal of the Acoustical Society of America*. October 1981, *70*(4), 985–995.

Songco, D. C., Allen, S. I., Plexico, P. S., & Morford, R.A. How computers talk to the blind. *IEEE Spectrum*, May 1980, 34–38.

Voiers, W. D. *Antecedents of speech acceptability*. Paper presented at the International Symposium on Measurements in Telecommunications, Lannion, France, October 1977.

Wiggins, R. H., & Brantingham, L. Three-chip system synthesizes human speech. *Electronics*, August 1978, *51*(18), 109–116.

Woodson, W. E., & Connover, D. W. *Human engineering guide for equipment designers*. Berkeley, CA: University of California Press, 1964.

9

Designing Chunks for Sequentially Displayed Information*

ALBERT N. BADRE
Georgia Institute of Technology
Atlanta, Georgia 30332

The objective of this study was to determine the significance of chunking in the design of sequentially presented information. Experiments were conducted to investigate: (a) the effect on recall accuracy of the sequential displaying of information chunks, and (b) the effects on recall accuracy and chunking characteristics of information presented on a display which itself is viewed as a member of a sequence of displays.

The general procedure was to utilize tactical scenarios to be reconstructed and copied under varying conditions of presentation and viewing by a select group of well-practiced subjects. The first experiment's hypothesis was that presenting tactical information incrementally and sequentially by meaningful chunks leads to greater information assimilation (measured by recall accuracy) than if the information is presented by a sequence of non-meaningful chunks. Another hypothesis was that the presentation order of meaningful chunks will have a significant effect on assimilation performance. Both hypotheses were supported by the data. The results suggest that not only is it important to present information in meaningfully organized chunks, but where the information cannot be presented all at once, information assimilation improves when the meaningful chunks are presented in a user-meaningful sequence.

The second experiment examined the effects of varying the placement of an information display in a sequential context of displays on chunking characteristics and recall accuracy. The findings showed no difference in recall accuracy under varying placement conditions. This result suggests that the viewer's recall strategy is to abstract information, and formulate a theme relevant to the entire sequence of displays, not only to the one to be reconstructed. Sequential context had no effect on chunk size. The availability of an invariance in the sequence of displays had a significant effect on the order in which symbols were reconstructed and on chunk content.

---

* This research was supported in part by Army Research Grant #MDA-903-78-0-04.

1. INTRODUCTION

Effective management of information is one of the major problems faced by decision makers. A decision maker has to gather, represent, process, assimilate, and use large amounts of information. In many cases, decisions have to be made in rapidly changing information environments. One way to cope with the problem of information management by decision makers is to provide them with on-line decision-aiding tools. There is a growing consensus, however, that such on-line aiding systems become usable when they are designed to be compatible with the user's information-processing capabilities and limitations (Badre, 1980).

A great deal of emphasis has been placed in the past on the visual factors of computer display design (Gould, 1968). Important work has been done to investigate the effects on performance and legibility of factors such as illumination levels, display density, code size and type, stroke width, and figure ground (Bennett, Chitlangia, & Pangrekar, 1977; Cahill & Carter, 1976; Hepler, 1976; Konz & Mohan, 1972; Vartabedian, 1971a, 1971b). More recently a growing emphasis is being placed on considering the information and language-processing aspects of computer interaction design (Stewart, 1976; Badre, 1980; Thomas & Carroll, 1981).

Designers of user-compatible on-line systems for decision processing need criteria for algorithms that search for, classify, and order the display of user-meaningful information chunks. The manner in which information should be organized and presented for effective processing in interactive decision tasks depends heavily on the decision maker's cognitive strategy for chunking the information on a display (Badre, 1982; Chase & Simon, 1973; Allen, 1982). The selection of meaningful chunks is likely to be impacted by contextual variables such as the perceived flow of information, the mode of presentation, e.g., textual or graphic, and the decision maker's functional objectives due to differences in training, perspective, or expertise.

In this study, we use tactical decision scenarios in combination with memory experiments to determine the effects of contextual variables on the interactive operator's recall performance of presented information. We assume that an operator's ability to recall displayed information is directly related to his/her ability to assimilate and use the information in rapid interactive decision situations.

Previous research focused on identifying characteristics of the informational chunks that novices and experts formulate when viewing displayed information about tactical situations (Badre, 1979; Badre, 1982). The main conclusion from that research was that meaningful chunks of information are identifiable for "coherent" information displays. For example, a chess position is more or less "coherent" if it comes out of a more or less well-played game. The size of the chunks as well as the frequency with which meaningful chunks were formulated were significantly greater for coherent than for non-coherent displays. It was also clear that the average chunk is composed of a number of relations between atomic units of information, e.g., unit designator symbols as in Figure 1. The underlying thesis of the earlier research was that the expert analyzes and processes the viewed displays in terms of well-formed structures (chunks), and that these structures provide the basis for selecting and evaluating the foregrounds of play or action. The information to which the problem solver attends on a given display constitutes the foreground of that display. The well-formed structures are the tools and elementary

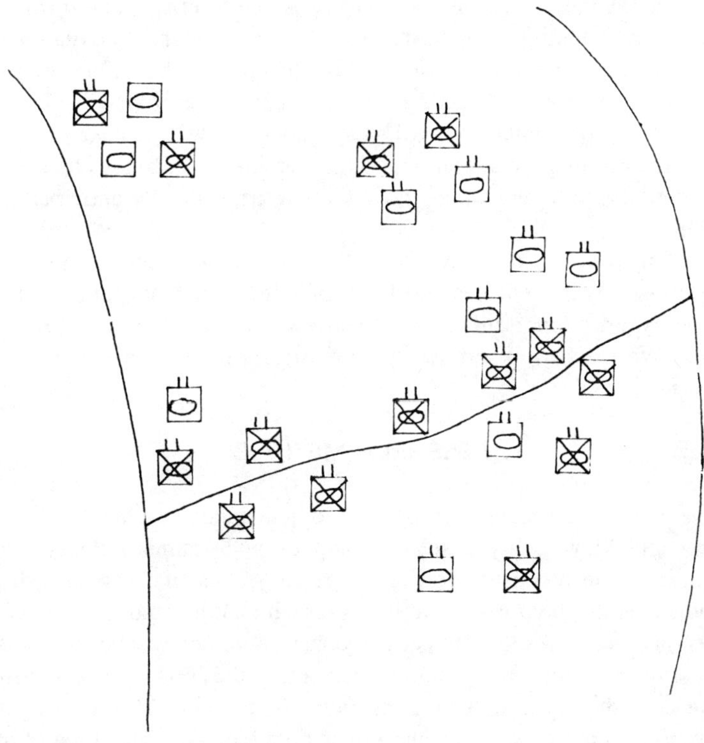

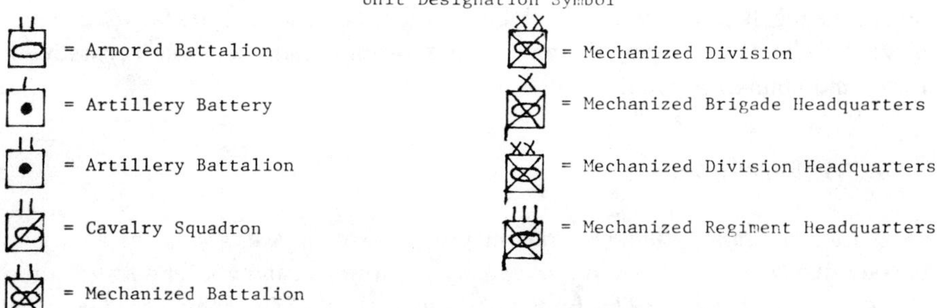

FIGURE 1. Tactical Display

vocabulary, possibly nonverbal, used in foreground perception and synthesis. When the constraint of sequential processing applies, as it may in the viewing and analysis of on-line displays, the expert problem solver is likely to process his/her information in an incremental predetermined order of meaningful chunks.

The chunking conjecture implies that the designer of displayed information should take into account the chunking rules identified in the user's practices and behaviors (Badre, 1982). But in order to relate the chunking conjecture more realistically to "interactive" situations, we must consider the effects of dynamic decision scenarios on chunking. It seems reasonable to assume that in dynamic situations (ones requiring a sequence of displays to convey the information), the underlying representation for selecting a set of chunks for one of the displays is not governed solely by or limited to the display under analysis. Rather, it stems as well from the given display's relations to the sequence of displays that were its immediate predecessors, as well as those that are immediately anticipated by it (Moses, 1982).

Accordingly, in addition to testing the sequential-presentation-of-chunks conjecture, this study investigates the effects of a line of play or action in a tactical game on the expert's representation as manifested in the chunks and perceived foregrounds. We ran two experiments to investigate these conjectures.

2. RESEARCH METHOD

Tactical scenarios were reconstructed and copied under varying conditions of presentation and viewing by a select group of well-trained subjects. In the simplest form of the reconstruction task, the subject is first shown a display. He/she studies the display for a fixed time after which it is removed and he/she is asked to reconstruct it. As the subject is reconstructing the display symbols, the experimenter records time and symbol placement. In the copying task, the subject is given the same display as in the reconstruction task and is asked to copy it on a blank diagram as rapidly and as accurately as possible. The same type of data is recorded here as in the reconstruction task. The data from both tasks is used to determine recall accuracy and chunk boundaries. These techniques are used to analyze the effects of varying the presentation of information on both accuracy of recall and chunking charcteristics.

2.1 Experiment I

Earlier findings indicate that in problem-solving situations, experts can encode displayed information more easily if shown meaningful chunks (Badre, 1979; Chase & Simon, 1973; Frey & Adesman, 1976). The conjecture here was that if meaningfully-chunked information is presented to the subject incrementally in a sequence familiar to him/her, it is likely to lead to greater information assimilation (measured by recall accuracy) than if the same information is presented either by a sequence of non-meaningful chunks or by meaningful chunks that are presented in an order different from that to which the subject is accustomed.

2.1.1 Subjects. Thirty-six volunteers, well-trained in the vocabulary of tactical decision making, were selected on the basis of experience to participate in the experiment.

2.1.2 Material. The material for this experiment consisted of three displays similar to the example in Figure 1. The symbols of each display were grouped into two sets. One set consisted of a meaningful collection of chunks and the other of chunks whose constituent symbols are not relatable in a meaningful way. The meaningfulness of a chunk and the order of its presentation were determined on the basis of previously completed research (Badre, 1982). The displays were presented to the subjects on film.

2.1.3 Design and Procedure. Comparisons were made among four presentation modes. These are:

1. the one-shot presentation of the complete display as shown in Figure 1;
2. the presentation of the display incrementally by chunks in an order established by previous research (Figure 2 provides an illustration of the various chunks);
3. the development of the display by chunks in reverse of the already established order; and
4. the development of the display incrementally by chunks whose constitutent symbols have no meaningful relationships.

Three tactical displays (S1, S2, and S3) were used, each in the four presentation modes described above. The subjects were randomly assigned to four groups of nine subjects. Each subject was shown the three displays in one of three of the four different presentation modes. The display presentation order was counterbalanced within each group. Table 1 shows how the presentation modes and the displays were distributed over the groups.

Displays were sequenced on a movie film in such a way that the three display presentations for a subject in a group could be processed in order by always skipping forward on the film, as in the following example for Groups 1 and 2.

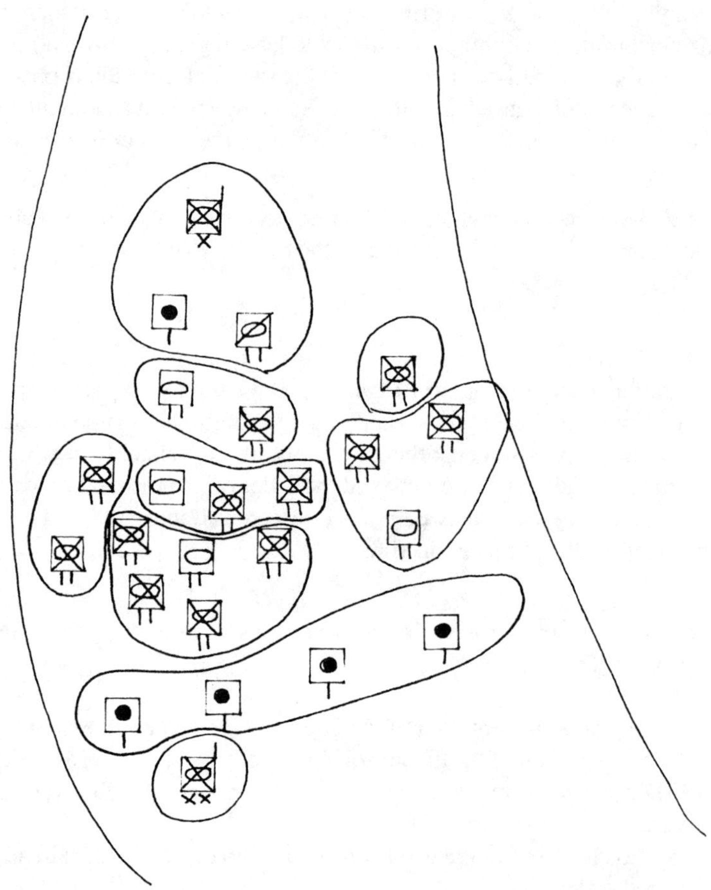

FIGURE 2. Display in Meaningful Chunks

TABLE 1
Summary of Displays × Groups of Subjects × Presentation Modes

		Subject Groupings			
		Group 1	Group 2	Group 3	Group 4
Tactical	S1	b	d	a	c
Displays	S2	a	c	d	b
	S3	d	b	c	a

a = one-shot display
b = correct sequencing of meaningful chunks
c = reverse sequencing of meaningful chunks
d = sequencing of non-meaningful chunks

$$
\text{Group 1} \begin{cases} \text{S1} & \text{b} \\ \text{S2} & \text{a} \\ \text{S3} & \text{d} \\ \text{S4} & \text{b} \\ \text{---} \\ \text{---} \end{cases}
\qquad
\text{Group 2} \begin{cases} \text{/S1d} \\ \text{S2c} \\ \text{S3b} \\ \text{---} \\ \text{---} \end{cases}
$$

Each subject was told that this is an experiment in information processing consisting of three trials. In each trial a tactical display would be shown briefly on film. The display may be shown all at once in one exposure or may be developed incrementally in a sequence of several film frames, each frame lasting between two and three seconds. After 18 seconds of viewing time, the display was removed, and the subject was asked to reconstruct it on a sheet of paper that has on it the outline of a display background. For reconstructing the display the subject used rubber stamps with the proper symbols. A pretest slide was used for practice. The reconstruction task was followed immediately by the copying task. Here the participant was asked to copy the symbols on each display as accurately and as rapidly as possible.

2.2 Experiment II

The theme of this experiment is that in tactical analysis situations, the underlying representation as manifested in chunking behaviors resides not in the single position under analysis, but in a time sequence of related and updated tactical scenarios. This theme leads to the following conjectures:

1. Chunking characteristics will differ if a tactical display is presented out of context under one condition and then presented as a member of a sequence of displays under another condition.
2. Given a coherent sequence of displays, one where the displays fall logically and realistically in the order presented, both accuracy of recall and chunking characteristics will not differ significantly as a function of the position in the sequence of the criterion display (the one to be recalled).
3. Given a random non-coherent sequence of displays, chunking characteristics and accuracy of recall will differ significantly as functions of the criterion display's position in the sequence.

   An experiment was designed where the display to be reconstructed was presented under different conditions of sequencing.

2.2.1 Subjects. Thirty-five volunteers who are experts in tactical decision making were used as subjects in this experiment. They were randomly assigned to five groups of equal size.

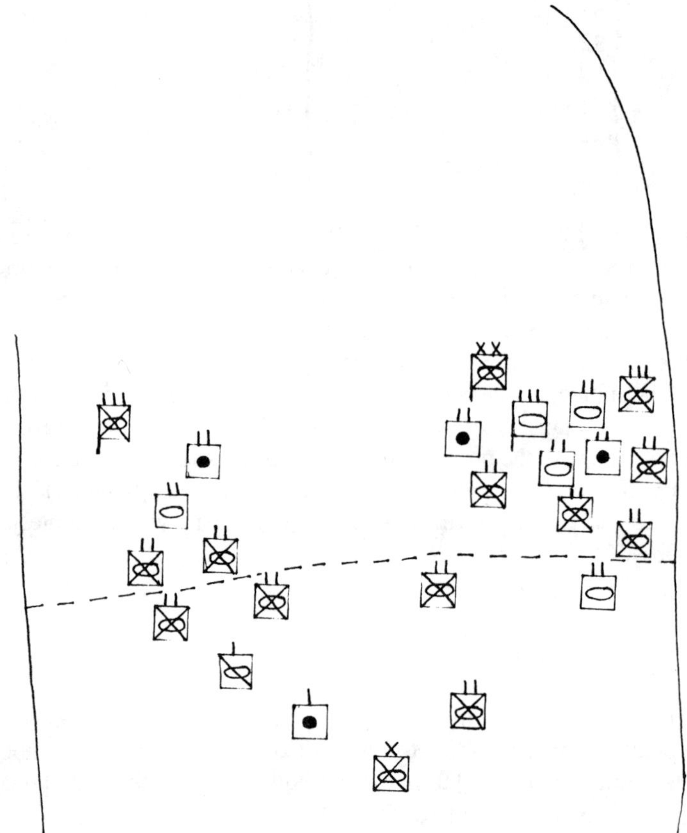

FIGURE 3. The Display-To-Be-Reconstructed − *F*

2.2.2 Material. The material for this experiment consisted of a sequence of nine distinct displays that follow each other in a realistic order of tactical action. The displays were presented to the subject in ordered sequences of five displays at a time. Figure 3 shows the criterion display—the one to be reconstructed.

2.2.3 Design and Procedure. The 35 subjects were randomly assigned to five equal groups and underwent the same procedure for the reconstruction and copying tasks as in Experiment I. Each group was associated exclusively with one of five presentation modes. The presentation modes differed on the position where the display-to-be-reconstructed F occurred in the sequence in Table 2. For example, the first sequence starts with display F and ends with display J.

Under each mode, the complete line of displays was presented on a rectangular cardboard at once and held in view for 45 seconds before being removed.

TABLE 2
Position of Criterion Display in Presentation Mode

Position of Criterion Display F
in Mode of Presentation Sequence

first	middle	last	alone	middle out of sequence
F	D	B	*F*	I
G	E	C		B
H	*F*	D		*F*
I	G	E		H
J	H	*F*		C

Then, immediately, a blank "map" was placed in the position where display F had occurred indicating that the subject should reconstruct F.

After both experiments were finished, 16 subjects were asked to perform the copying task. Here the subject was presented with a display diagram that was placed on a table behind him/her and a blank sheet placed on the table in front of him/her. The subject was told to copy the symbols from the diagram onto the sheet by placing them in the appropriate locations quickly and accurately. Each subject copied only one of the four displays used in both experiments. Thus, each display was copied by four different subjects.

3. RESULTS AND DISCUSSION

3.1 Data Collection and Analysis

For both of the experiments, the data collection was the same. There were essentially two kinds of data collected for both the reconstruction and copying tasks. These were symbol placement times and order-of-symbol placements. In the first case, one of two experimenters recorded the times of the placement of symbols via a cassette tape recorder. This same experimenter also kept time for the 5-second presentation in the reconstruction task. For the copying task, the experimenter recorded the symbol placement times and the times for the beginning and end of a glance at the diagram from which the subject was copying. The second type of data collected was the order in which the symbols were placed on the blank diagram. This data was collected by the second experimenter. He recorded the order by using a blank diagram and writing the original number in the location corresponding to that used by the subject to place the symbol.

There were four measurement criteria associated with the data described above. These are: (a) the number of accurately placed symbols (a symbol is placed accurately if both its value and location are correct); (b) the order of placement of accurately placed symbols; (c) the inter-placement times between symbols (IPT); and (d) the within-glance symbol identification and time counts for the copying task.

Several assumptions are made. First, in the copying task, it is assumed that successive glances to the diagram from which symbols are being copied define the boundary of chunks. That is, the symbols that are placed on the response diagram between two glances to the stimulus diagram are referred to as the within-glance symbols and considered to constitute a chunk. Second, the average IPT is computed for the within-glance symbols of each subject and used to define the chunk boundaries in the reconstruction task. Symbols placed at or below the computed IPT were assumed to belong to the same chunk; those falling above the computed IPT were considered to come from two different chunks, hence defining a chunk-boundary in the reconstruction task. Finally, the content characteristics of chunks in the reconstruction task are compared with that in the copying task. Chunk comparison, for both groups of subjects, is made on data characteristics such as the size of a chunk, IPT distributions, within-chunk symbol relations and patterns, and order-of-chunk placements on diagrams. The order of placement is an indication of the importance of the chunk.

3.2 Presenting Chunked Information and the Accuracy of Recall

The times which fall within a glance in the copying task were averaged to-gether to provide the IPT. The average IPT for determining chunk boundaries was computed at 1.14 seconds. The overall percent accuracy for comparing the various modes of presenting chunked information is shown in Table 3. An analysis of variance yields a significant main effect for presentation modes with $F(3, 92) = 8.317$, $p < 0.01$.

The means show that presenting information sequentially and incrementally by meaningful chunks results in higher assimilation than if the same information is presented in a sequence of non-meaningful chunks. A t-test comparing the two

TABLE 3
The Overall % Accuracy and Standard Deviations
for the Four Modes of Presentation

Code	Value Label	%	Standard Deviation
1	Meaningful chunks	41.32	.1958
2	All-at-once	50.33	.1812
3	Non-meaningful chunks	31.44	.1453
4	Reverse chunks	29.33	.1331

means yields significance at $t(44) = 1.93$, $p < 0.06$. Also the means show that the correct sequence of meaningful chunks is significantly superior to the mode where the meaningful chunks are presented in the reverse order of reconstruction at $t(47) = 2.52$, $p < 0.05$. A t-test yielded nonsignificant difference between the mode where the chunks were presented all at once and that where they were presented incrementally in meaningful chunks, at $p > 0.1$.

These results suggest that it is important to present information in meaningful chunks, but when information cannot be presented all at once, information assimilation increases when the meaningful chunks are presented in a user-practiced sequence. The presentation of non-meaningful chunks resulted in higher accuracies than the presentation of meaningful chunks in the reverse-reconstruction order. This can be explained by looking at the average number of symbols common to the three presentation modes at successive chunk increments. Table 4 shows that the non-meaningful and meaningful modes have many more symbols in common (9.67 for the three displays) after five increments, than do the meaningful and reverse conditions (three for the three displays). Hence, the displays of the non-meaningful mode contain meaningful chunks for a longer period of time than do the displays of the reverse mode. The availability of meaningful chunks for longer viewing and assimilation durations may have led to higher accuracy performance for the non-meaningful condition than for the reverse one.

TABLE 4
Mean Number of Identical Symbols That are Present on the Different Displays for Each Increment of Chunks

Number of chunks on display	A	B
1	.33	0
2	1	0
3	2	0
4	5.33	0
5	9.67	3
6	11.33	8
7	15.67	14.33
8	21.33	20
9	24	24

A = Average number of symbols common to the non-meaningful and meaningful modes.
B = Average number of symbols common to the meaningful and reverse modes.

3.2 The Effects of Sequential Context on Chunking Characteristics and Accuracy of Recall

To determine the effects of sequential context on recall accuracy and characteristics of the reconstructed chunks, several comparisons between contextual modes were made.

3.2.1 Accuracy. The extent to which the association between displays is well or loosely structured may be determined by comparing the conditions where the display-to-be-reconstructed is placed at either the start, the middle, or the end of a sequence of displays. The assumption is that if there is a significant difference on accuracy between the three conditions, then a memory differential and interference effect may exist. If, on the other hand, there is no significant difference between the three conditions, then it may be suggested that the viewer is processing the salient features—the foreground of the entire sequence and not simply the foreground of individual displays. The analysis yields a nonsignificant difference at $F(2, 16) = 0.0918$, $p > 0.1$, hence supporting the suggestion that the viewer is encoding a foreground of the entire sequence of displays.

A t-test comparison on accuracy of reconstruction between a non-coherent sequence and a coherent one yielded non-significance with $t(11) = 1.42$, $p > 0.1$. It is important to note that both sequences contained a subset of symbols that remained constant, invariant, from one display to the next. When the subjects reconstructed a display, the recalled symbols were mostly the invariant ones. This result suggests that the constant features of an information display may be central to the process of assimilating the information. It may also be true that the availability of constant features may have an effect on how information is chunked.

3.2.2 Chunk Characteristics. In order to determine the effects of providing invariances in sequential context on various chunking characteristics, a comparative analysis was made between the condition where the display-to-be-reconstructed is presented alone, and the condition where it is presented in the context of a coherent sequence of displays. The coherent sequence of displays contains a subset of constant symbols, ones that do not change from one display to the next.

The reconstructed displays were segmented into chunks using the inter-placement time measure. The average number of chunks per display for all modes of presentation was 7.46. An analysis of variance yielded a nonsignificant difference at $F(4, 27) = 1.122$, $p > 0.1$.

In order to determine whether sequential context has an effect on the size of a chunk, a comparative analysis was made for the five modes of presentation. Two different units of chunk content were used to determine size: symbols and tactical relations between symbols. An analysis of variance for the means in Table 5

TABLE 5
Means and Standard Deviations of the Number
of Symbols per Chunk for the Five Modes
of Contextual Presentations

Contextual Position of Criterion Scenario	Mean	Standard Deviation
First of 5	2.2000	1.4405
Middle of 5	2.0000	.7071
Last of 5	2.9286	.9759
Middle of 5 (unstructured sequence)	2.8000	.8367
Single (alone)	1.8571	.5563

yielded nonsignificance among the five modes with respect to the number of symbols per chunk, with $F(4, 26) = 1.802, p > 0.1$.

This result is consistent with the no-difference results over modes for accuracy of reconstruction and number of chunks per display (Badre, 1982). A similar analysis for number of relations per chunk yielded a nonsignificant $F(4, 26) = .6905, p > 1.0$. The average number of relations per chunk is 1.64.

An indirect way of determining the effects of an invariance in a sequence of displays on assimilation is to examine the order in which symbols are reconstructed. Earlier research showed that when a subject reconstructs a display, when the display is presented out of context, he/she invariably places a significantly greater number of red symbols in the first two reconstructed chunks. In Figure 1, red symbols are those placed in the upper half of the display. In most tactical situations, a red symbol designates a unit belonging to the opponent, a blue symbol designates a unit that is friendly. A likely hypothesis is that an invariance provides a basis for extracting the foreground of displayed information and thus is a focal point of information assimilation. Hence, if given a group of invariant symbols, all of which are blue, it is highly probable that the participant would begin by reconstructing more blue symbols than red ones in the first two reconstructed chunks. In comparing the color of reconstructed symbols for the first two chunks, between the sequential context mode and the mode where the display-to-be-reconstructed was presented alone, it was clear that the availability of an invariance made a significant difference. For the mode with no invariance, the percent of red symbols in the first reconstructed chunk is 78%; for the first two reconstructed chunks, it is 43%. This result replicates earlier findings (Badre, 1982). For the presentation mode where an invariance is provided, the percent of red symbols in the first two chunks is zero. The difference between the two modes for the mean number of red symbols over the first two reconstructed chunks is significant at $t(12) = 2.12$, $p < 0.05$. This finding is a clear indication that the availability of invariances in information displays may have an effect on chunk content and the way information is perceived and organized.

4. THE IMPLICATION FOR DESIGNERS

The results show that chunking displayed information in a way that is meaningful to the user leads to easier assimilation than if the same information is clustered into an equal number of non-meaningful chunks. Also, the presentation order of meaningful chunks has an effect on ease of assimilation. If the chunks are presented sequentially in the order with which the user is familiar, then the information is assimilated faster than if the same user-meaningful chunks are presented in the reverse of the user-familiar order. This is true even though the logical connectedness and coherence of the information is preserved. This result implies that in the design of interactive applications where the information requires several displays, the information segmentation and presentation order should not be taken for granted. The user-information processing and management habits should be studied and understood.

The results of the second experiment suggest that the viewer's assimilation strategy is to formulate a theme relevant to the entire sequence of displays. The theme itself, as well as what information is chunked and the order of chunk formulation seem to be influenced by a subset of information that is constant, invariant, throughout the entire sequence. This invariance seems to provide a point of focus for remembering the theme. This result implies that designers should make salient the invariant features of displayed information to aid the novice in synthesizing the details of a sequence of displays, e.g., providing thematic titles on screens.

The subjects used in this study were experts in the language of military displays. They had well-established and trained habits of viewing and organizing the information on tactical displays. Both the chunk contents and the order in which the chunks were presented were important for their ability to assimilate the information. When the information was presented and organized in ways foreign to their training and experience, e.g., non-meaningful chunks or reverse-of-reconstruction order, they performed poorly. When the displays were organized to match the chunking strategies to which the subjects were accustomed, then performance improved. This result implies that designing usable information on displays means taking into account the chunking rules and strategies identified in the user's practices and behaviors.

Often the tendency for developers and designers is to design information displays in ways that are usable by them. They consider themselves the role models for interacting with information. They disregard the possibility that the way they use the information as developers may be quite different from the way the end user learned to view and use the same information.

The designer should understand that the same information can be chunked and presented in various ways. But only a select few chunking and presentation strategies are readily compatible with the user's habits and practices. It is imperative then that before designing information for interactive displays in a given user environment, the designer should study and determine the rules governing the information-processing and organizing habits of the end user.

5. REFERENCES

Allen, R.B. Cognitive factors in human interaction with computers. In A. Badre & B. Shneiderman (Eds.), *Directions in human/computer interaction*. Norwood, NJ: Ablex Publishing Corp., 1982.

Badre, A.N. Human cognitive factors in front-end interaction. (Reprinted from *Proceedings of CompCon,* September 1980.)

Badre, A.N. *Selecting and representing information structures for tactical decision systems.* (ARI Tech. Rep. DAHC 19-77-G-0022.) 1979.

Badre, A.N. Selecting and representing information structures for visual presentation. *IEEE Transaction on Man, Systems and Cybernetics,* January 1982.

Bennett, C.A., Chitlangia, A., & Pangrekar, A. Illumination levels and performance of practical visual tasks. *Proceedings of the Human Factors Society, 21st Annual Mtg.,* 1977, 322–325.

Cahill, M.C., & Carter, R.C., Jr., Color code size for searching displays of different density. *Human Factors,* 1976, *18,* 273–280.

Chase, W.G., & Simon, H.A. The mind's eye in chess. In W. G. Chase (Ed.), *Visual information processing.* New York: Academic Press, 1973.

Frey, P.W., & Adesman, P. Recall memory for visually presented chess positions. *Memory and Cognition,* 1976, *4* (5).

Gould, J.D. Visual factors in the design of computer-controlled CRT displays. *Human Factors,* 1968, *10,* 359–376.

Hepler, S.P. Continuous versus intermittent display of information. *Human Factors,* April 1976, *18.*

Knapp, B. G., Gellman, L. H., & Moses, F. Information highlighting on complex displays. In A. Badre and B. Shneiderman (Eds.), *Directions in human/computer interaction.* Norwood, NJ: Ablex Publishing Corp., 1982.

Konz, S., & Mohan, R. The effect of illumination level, stroke width and figure ground on legibility of name. numbers. *Proceedings of the Human Factors Society,* 1972.

Thomas, J.C., & Carroll, J.M. Human factors in communication. *IBM Systems Journal,* 1981, *20* (2).

Vartabedian, A.G. Legibility of symbols on CRT displays. *Applied Ergonomics,* September 1971a.

Vartabedian, A.G. The effects of letter size, case, and generation method on CRT display search time. *Human Factors,* August 1971b, *13.*

10

Information Highlighting on Complex Displays*

BEVERLY G. KNAPP
& FRANKLIN L. MOSES
US Army Research Institute

and

LEON H. GELLMAN
Sarah Lawrence College

Recent work in the area of highlighting complex information on computer-driven displays is presented. Emphasis is on two categories of formatting techniques to enhance information assimilation: coding— represent information to maximize symbol differentiation, and sequencing—show information in segments or groups over time. Using a battlefield situation map as a prototype complex display, examples of coding and sequencing information are discussed. Coding examples include double cue coding and color coding; sequencing is represented by windowing, zooming, and selective call-up of aggregated information. The paper provides a list of preliminary guidelines for highlighting information on complex displays.

Computer-driven displays are valued for the sheer volume and speed of information that can be presented. Volume and speed of information, however, have also created new problems. If crucial information is not easy to extract clearly and quickly, then the benefits of automation can be lost. Guidelines are needed for highlighting or sequencing information on complex displays so that its format does not hinder communication value.

Much research attention concerning display formats has focused on parameters such as character legibility, symbol discriminability, operator viewing angle, etc. In fact, recent reviews of research (e.g., Meister & Sullivan, 1969; Christ, 1975; Engel & Granda, 1975; Snyder & Maddox, 1978) offer guidelines for a variety of display parameters (Table 1).

---

* The views expressed in this paper are those of the authors and do not necessarily reflect the views of the US Army or of the US Department of Defense.

TABLE 1

Display size	Type of character
Character size	Use of color
Brightness/contrast	Character/background relationship
	(light on dark, dark on light)
Resolution	Number of characters presented
Viewing distance	Frame (change) rate
Viewing angle	

Without question, such parameters are key features for defining display characteristics. Yet, there is a gap between defining character size, resolution, color, etc., and providing a well-formatted display of information. A display is more than just a collection of static elements; it must use aggregated elements to support a dynamic user/display interaction. The current paper discusses how display formats can be chosen systematically for "decluttering" displays and clearly presenting clusters of information.

COMPLEX BATTLEFIELD DISPLAYS

The need for clear formats is particularly well illustrated by requirements of battlefield displays. Current US Army requirements call for the development of automated graphic displays to provide an overview of the battlefield for situational assessment and decision making. Typically, these displays consist of a topographic map with overlays symbolizing battlefield events. Figure 1 shows a simplified example. The challenge is to clearly format large amounts of changing data as it becomes available. For example, the operator in a battlefield automated system may be called upon to monitor inputs from a particular source, targets of a particular type, a particular attribute of complex stimuli, and outputs in a particular category. The goal with displays is to look for patterns and trends as well as singular discrete events where events often are inferred rather than directly sensed (Swets, 1976). The success of this interaction between display terminal and operator depends on how displayed information is partitioned, abstracted, or summarized in a form compatible with the operator's information needs. In summary, while basic specifications (Table 1) must be an integral part of display formatting, these are only the beginnings of issues for operator needs.

The Clutter Problem

As the operator aggregates information, a particularly troublesome problem is that displays appear too "cluttered" for effective comprehension. In essence, an automated display seems cluttered because it lacks order, or contains unnecessary elements. On a battlefield display, for example, unit symbols, terrain fea-

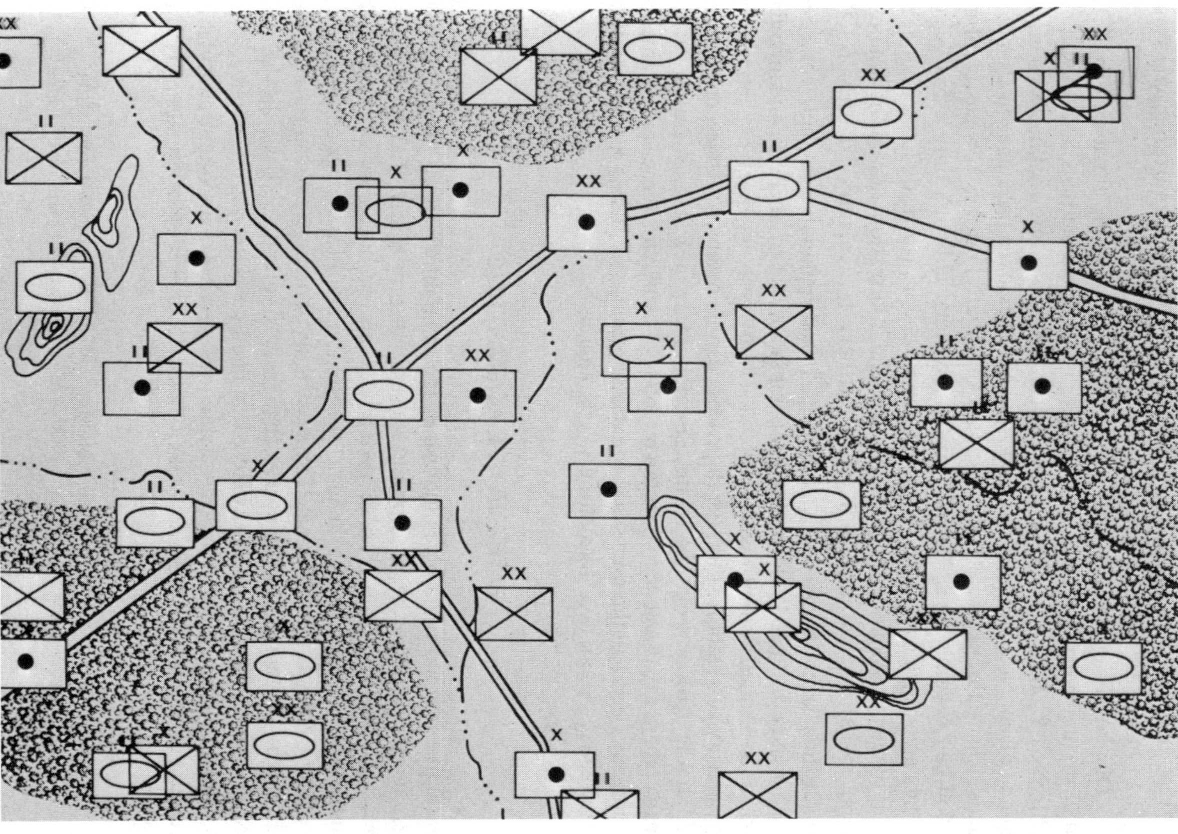

FIGURE 1. Simplified battlefield situation map display

197

sary elements. On a battlefield display, for example, unit symbols, terrain features, etc., can densely populate the limited visual display area. In short, when information updates occur and tactical situations change, then overlapping symbols and battlefield features accumulate.

Displayed information must be presented such that the situation is readily and easily understandable. Information that is most relevant to a task should be highlighted to stand out from less relevant information. Information to be highlighted, of course, necessarily depends on changes over time and on user demands. The fictitious battlefield situation map (Figure 1) shows how easily information can become "lost" on a display. Note that although only a few basic features are shown on this very stripped-down version, it already has a high density of elements. It shows only the most essential information: terrain (mountains, rivers, roads, forests); the unit types (artillery ⊡, infantry ⊠, armor ⊟); and the unit sizes (division, brigade, and battalion). Yet, consider the time and effort needed to compare the number of armor, artillery, and infantry units, even on such a simplified display. Research has shown that if a display has a high density of symbols, this search process can be extensive, time-consuming, and fatiguing (Bloomfield, Beckwith, Emerick, Marmurek, Tei, & Traub, 1979). Clearly, there exists a need to develop methods of displaying these same elements of information so that the user's task in extracting and assimilating the information is facilitated. Alternative formatting is one approach.

ALTERNATIVE FORMATTING

Alternative formatting involves the use of the same information, the same symbols, and the same hardware to produce a set of display screens having different perceptual characteristics. These differences are produced by varying such graphic dimensions as color, flashing, flagging, magnification, space, and time. Alternative formats are thus created by system manipulations, and an information display may be improved without data changes. By building on basic display and perception research, the US Army Research Institute (ARI) has investigat ed two categories of alternative formatting techniques which are important and viable methods for facilitating the user's task of extracting needed information from a graphic display:

1. *Coding:* represent information to maximize symbol differentiation.
2. *Sequencing:* show information in segments or groups over time.

Coding and Sequencing are efficacious methods of alternative formatting for high-density displays. However, multiple display screens and reduced detail displays also have potential. Multiple displays have the problem of added hardware expense but they provide a real increase in display area. Reduced detail techniques could be viable (e.g., Granda, 1976) although user surveys indicate a demand for more displayed information as opposed to less, especially in battlefield situations (Ciccone, Samet, & Channon, 1979; Landee, Samet, & Foley, 1979; Landee, Samet, & Gellman, 1980). Therefore, key advantages of using information coding and sequencing as display formatting techniques are that they minimize added hardware costs and decreases in user performance.

Coding

Coding helps to highlight displayed information by making symbols "stand out." This is accomplished by emphasizing particular symbols, elements, or features within the display with an auxiliary code. All information remains on the display, but critical dimensions are emphasized or redundantly portrayed. A literature review by Hemingway, Kubala, & Chastain (1979), has identified as many as 16 potential codes for highlighting information within an automated display. Among the graphic supplements which emphasize critical data are color, alphanumerics, shape, intensity, flashing, numerosity, and orientation. When these codes are combined, an enormous potential exists for highlighting data without the addition of new symbols. The coding techniques are based upon the perceptual principles of "similarity." For example, features which are similar will tend to be seen as one unit or "figure"; the other features are seen as background (Koffka, 1935). Coding of symbols emphasizes their figures and makes them easily distinguishable. Since "figures" are known to stand out from the "ground," they are more impressive, and more apt to be remembered (Hochberg, 1968). These qualities allow more effective user extraction of symbolic information from the display.

Single Versus Double-Coding Techniques. Vicino, Andrews, and Ringel (1965) investigated codes which are very efficient for displaying a change or update in specific information. Their research showed the value of double-cue or redundant coding, where information already on the display is reinforced with a conspicuous code. Examples of the coding techniques used in the single-cue and double-cue updates are shown in Figure 2, with the heavy outline used to distinguish codes.

Participants were presented successive pairs of slides and asked to detect changes. In the first slide, symbols were randomly positioned on a map, followed by a second slide where symbols were added, removed, or repositioned from the first. Data from single-cue and double-cue coding were compared with unaided performance (no coding). For all codes, the letter "N" represented an added symbol, dashed lines were used for removed symbols, and arrows for repositioned symbols. In general, increasing either the amount of information presented or the amount of updating resulted in degraded extraction and assimilation performance.

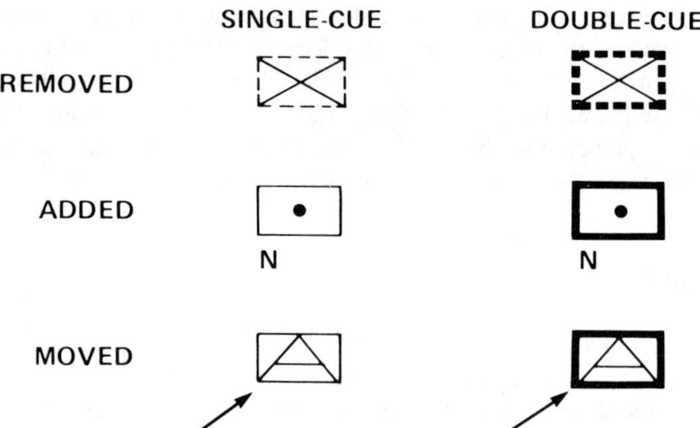

FIGURE 2. Examples of single cue and double cue coding

Double-cue coding completely prevented the degrading effect of increased amounts of information.

The power of double-cue coding is evident in Figure 3, which presents the situation map (from Figure 1) with double-cue coded unit symbols. The code is the heavy line used to emphasize the military unit types. There is no more or less information here: rather there are two ways of identifying the units. Clearly, the double code was very useful to maximize the saliency of symbols and to make them stand out together.

Color coding. How to use color as a coding technique is controversial. Some research evidence favors its use as a non-redundant display code for identification tasks, while other evidence recommends limiting color use to aiding detection or attracting attention. Krebs, Wolf, and Sandvig (1978) present both points of view in a comprehensive overview.

Using the battlefield display as a research vehicle, Sidorsky (1980)compared the use of color, shape, and alphanumeric codes. With the task of counting military unit types (e.g., artillery, infantry, and armor), Sidorsky found that color is a compelling way to specify salient information, provided the color-coded information is at the primary or first level of user interest (Figure 4). However, the color coding of unit types is of little help if the task of the analyst is to determine the number of units of different size (e.g., divisions, brigades, and battalions). Figure 5 illustrates the advantage of using color coding to indicate unit size when it is the primary information sought by the user. Data from counting tasks in Sidorsky's research based on comparisons among color, shape, and alphanumeric coding of symbols, show that color substantially reduces processing times. When there are doubts about how to assign redundant color coding in a system, color should be placed under user control by means of function keys.

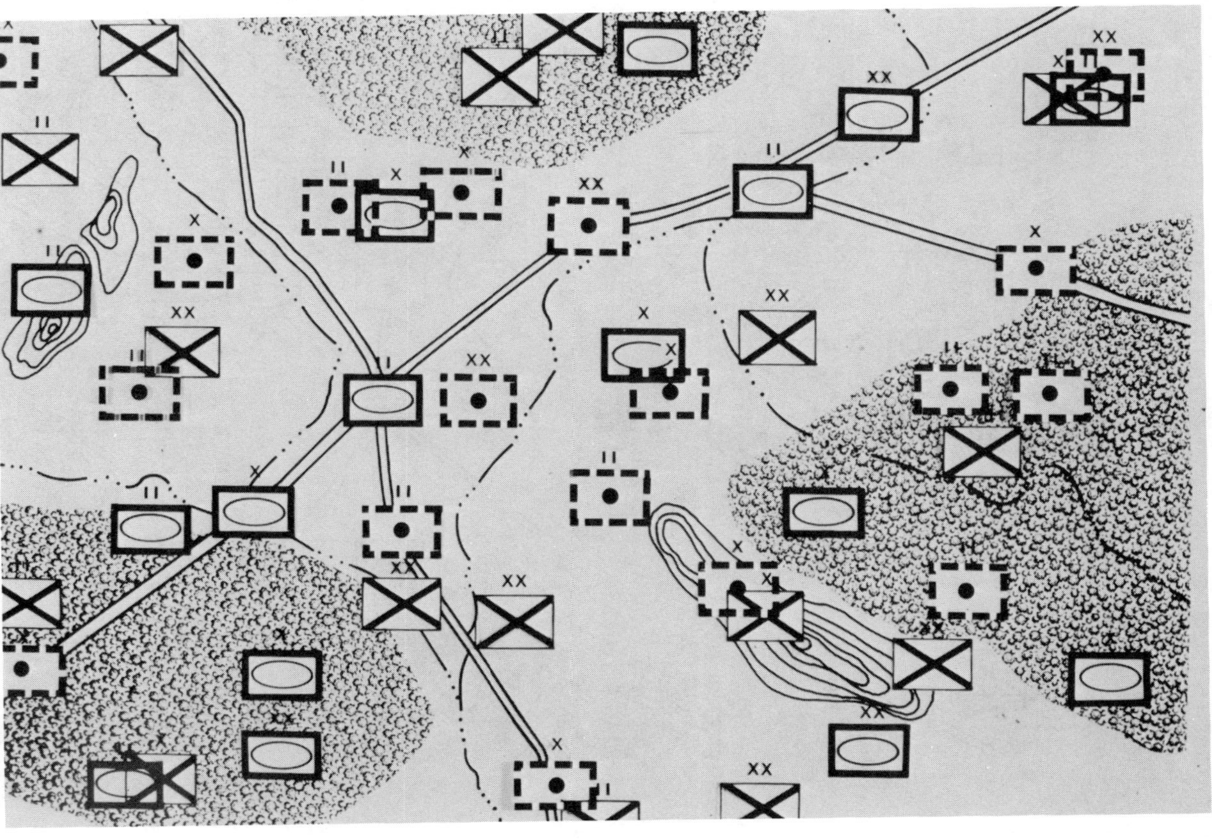

FIGURE 3. Battlefield situation display with double cue coding

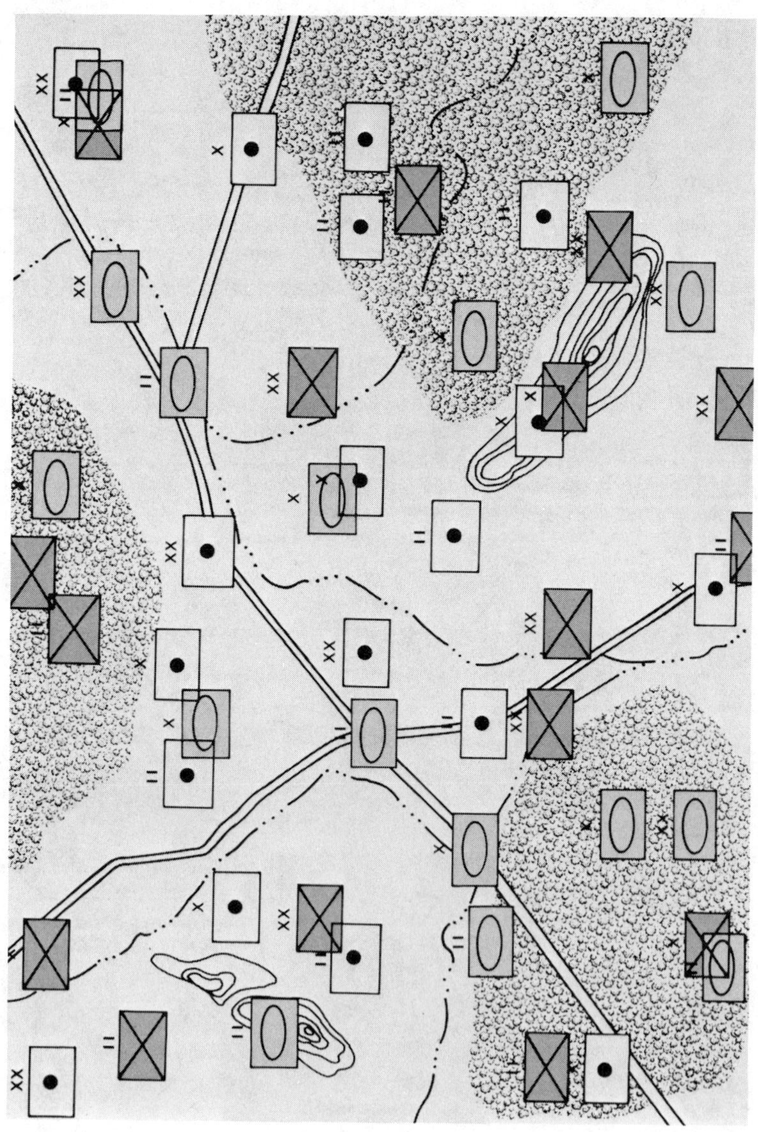

FIGURE 4. Color coding (shown in black & white) by unit type

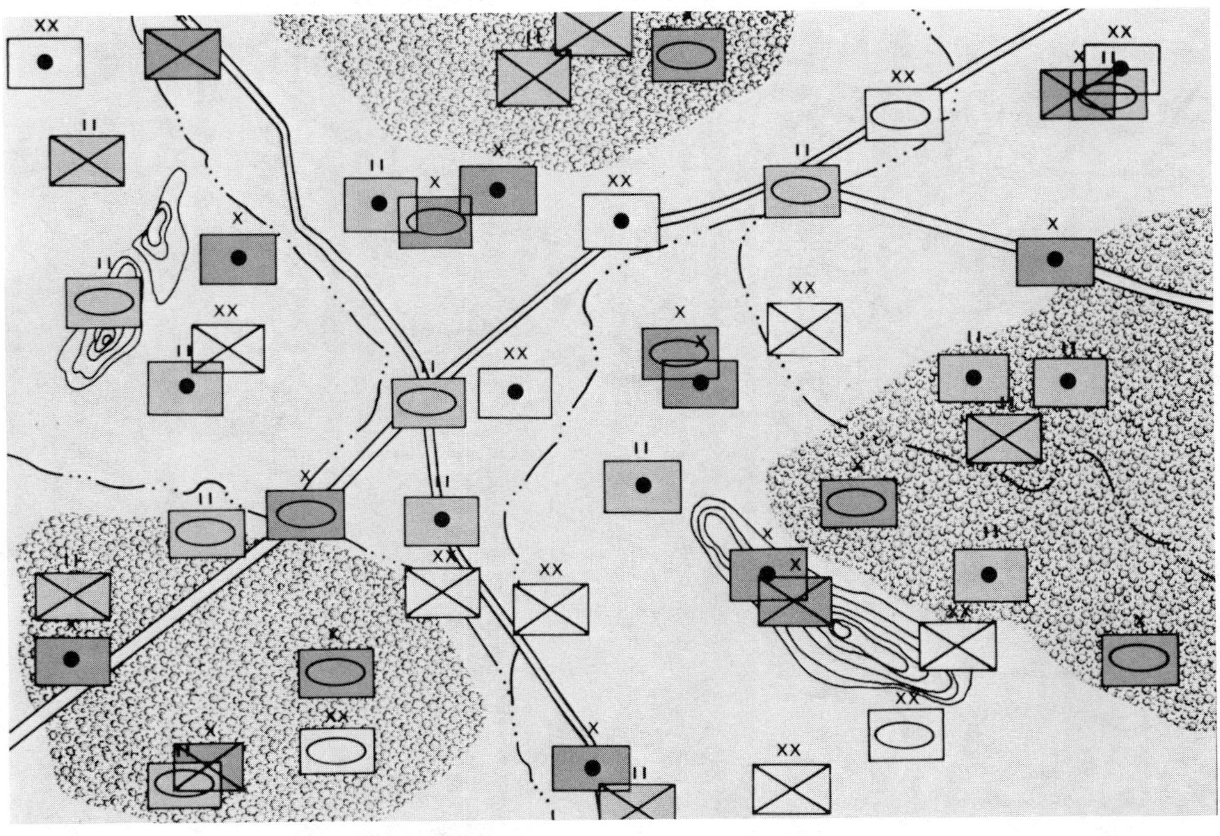

FIGURE 5. Color coding (shown in black & white) by unit size

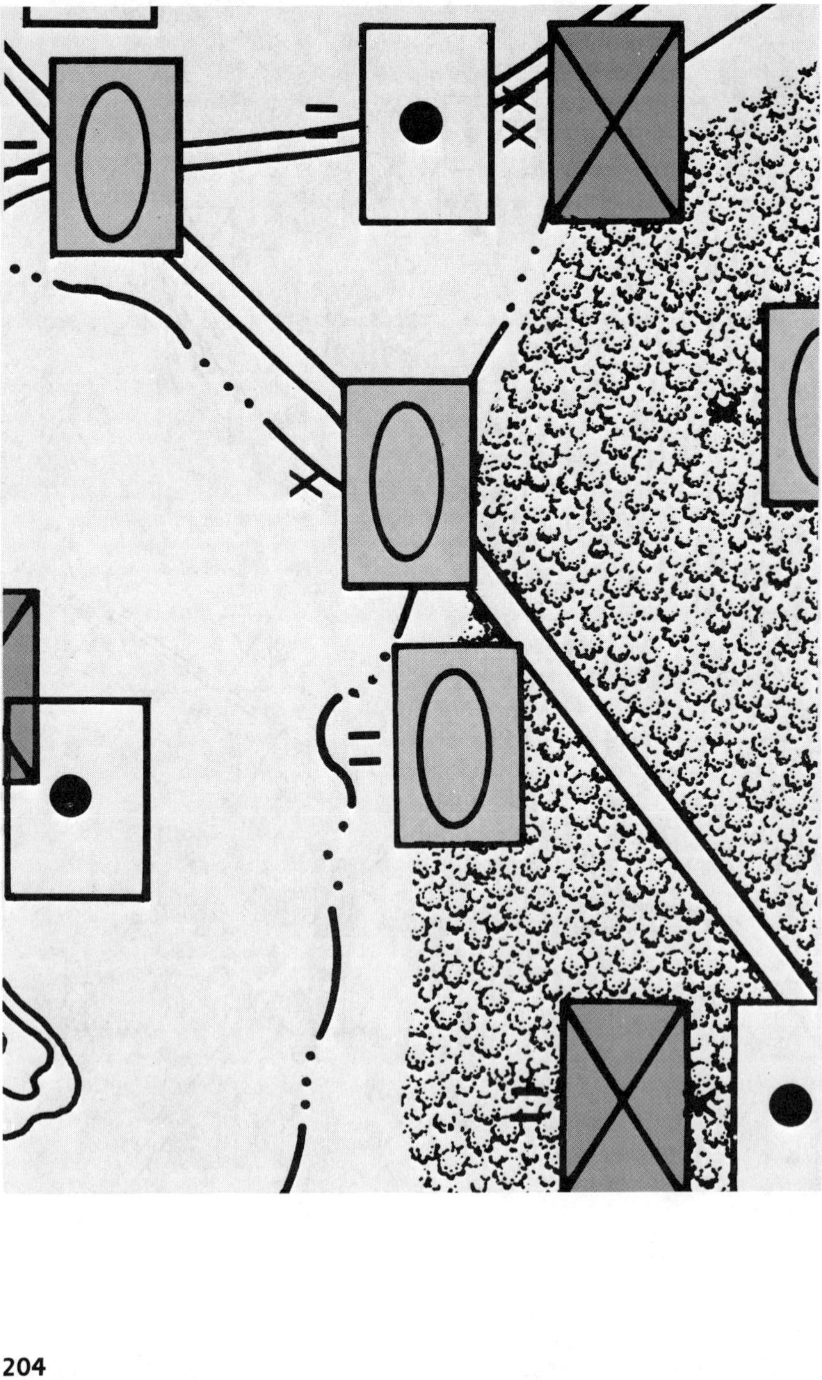

FIGURE 6. Enlarged segment of battlefield situation map

Sequencing

Sequential presentation reduces the number of symbols on the display at any one time by displaying segments of an entire display area over time. Figure 6 shows one segment of the sample map from Figure 1 and illustrates how magnification can reduce the density of information. Such sequential presentation of information allows assimilation of component parts, and increases the amount of detail possible per screen.

Display Windows. If the situation map is segmented as in Figure 7, then the display screen may be considered as a window through which the overall situation

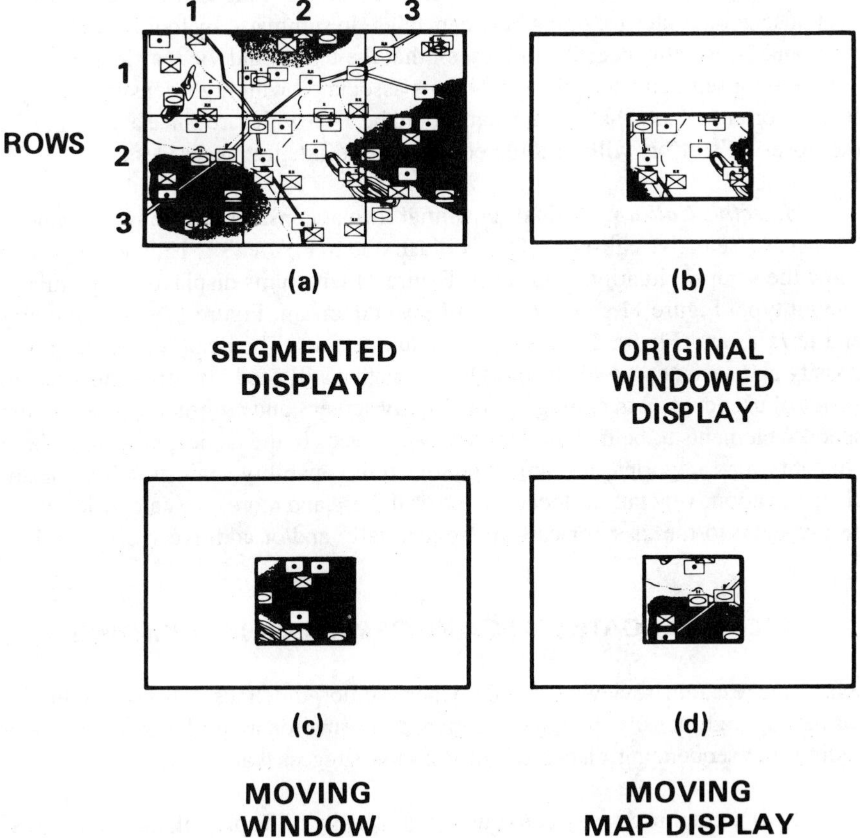

FIGURE 7. Segmented battlefield situation map for sequential viewing

map is viewed. Granda (1978) showed that the window can appear to move over the map or the map can appear to move under the window with no critical effect on performance. However, such sequential displays require users to depend upon their ability to integrate information over time. Thus, an important formatting question concerns whether to display segments of an entire situation map by scanning or by sequentially presenting discrete (i.e., static) views. Intuitively, the scanning method might seem to be the most flexible and promising for map displays. The user, however, might be served more quickly, less expensively, and equally well by sequential discrete displays. In an ARI experiment by Moses and Maisano (1979), military officers had to use map displays to find the quickest route from one city to another, depending on terrain, road types, etc. Sequential discrete views with overlaps of around 25% produced better results than continuous scanning methods of map presentation. It took less time to use discrete views to choose good routes for going between cities. In summary, by looking at Figures 8, 9, and 10 in sequence, it can be seen that the user could view a situation with overlapping segments to reduce problems associated with high-density displays. It is important to note that this format should be used with care since a large part of the observer's effort will be directed toward the integration process.

Selective Call-up. A final sequential format considered here is a technique known as "selective call-up." This is illustrated in Figures 11, 12, and 13, which show the sample situation map (from Figure 1) with units displayed sequentially by unit type. Figure 11 shows only artillery and terrain, Figure 12 shows infantry and terrain, and Figure 13 shows armor and terrain. The displays result in less density and easier comprehension. Questions that still need investigation relate to issues of user control in calling up the display screens and criteria for determining specific elements to be included on any one screen. In the tactical situation, these elements, or categories, are unit function, unit capability, unit mobility, obstacles, elevation, vegetation, location, etc. All these and more data categories could be available to the user for call-up, sequentially and/or additively displayed.

RECOMMENDATIONS FOR INFORMATION HIGHLIGHTING

Based on research reviewed in this paper, some guidelines can be offered for formatting high-density displays to enhance information assimilation. Data about coding and sequencing displayed information suggest that:

1. If a graphic display contains a high density of display elements, particularly overlapping and continuously updated features, then segregate the information using a coding or sequencing technique.

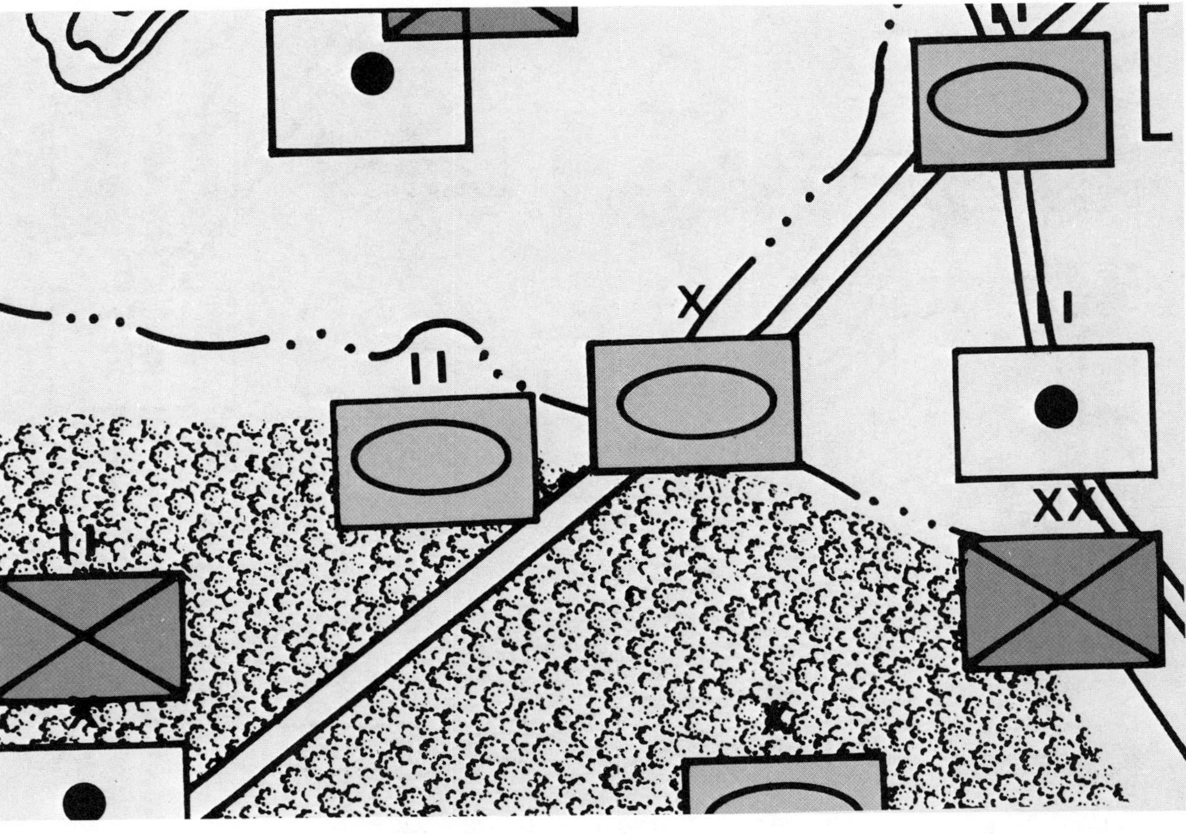

FIGURE 8. Enlarged segment of battlefield map

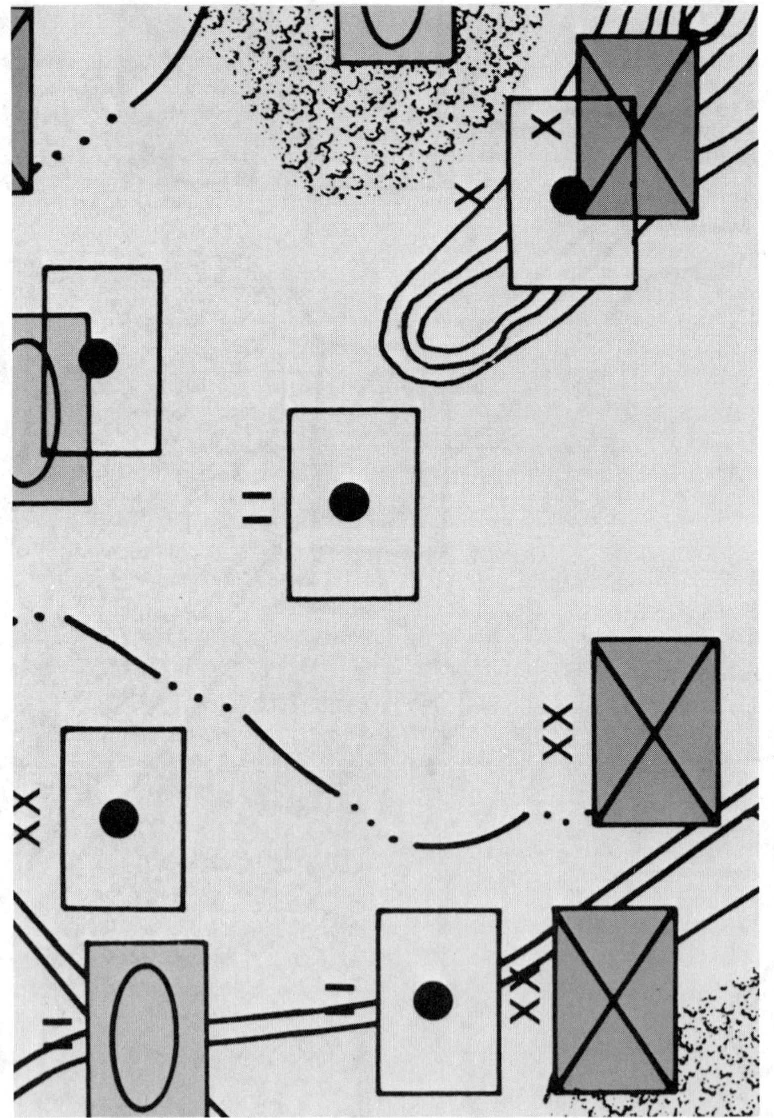

FIGURE 9. Enlarged segment of battlefield situation map with 25% overlap of Figure 8

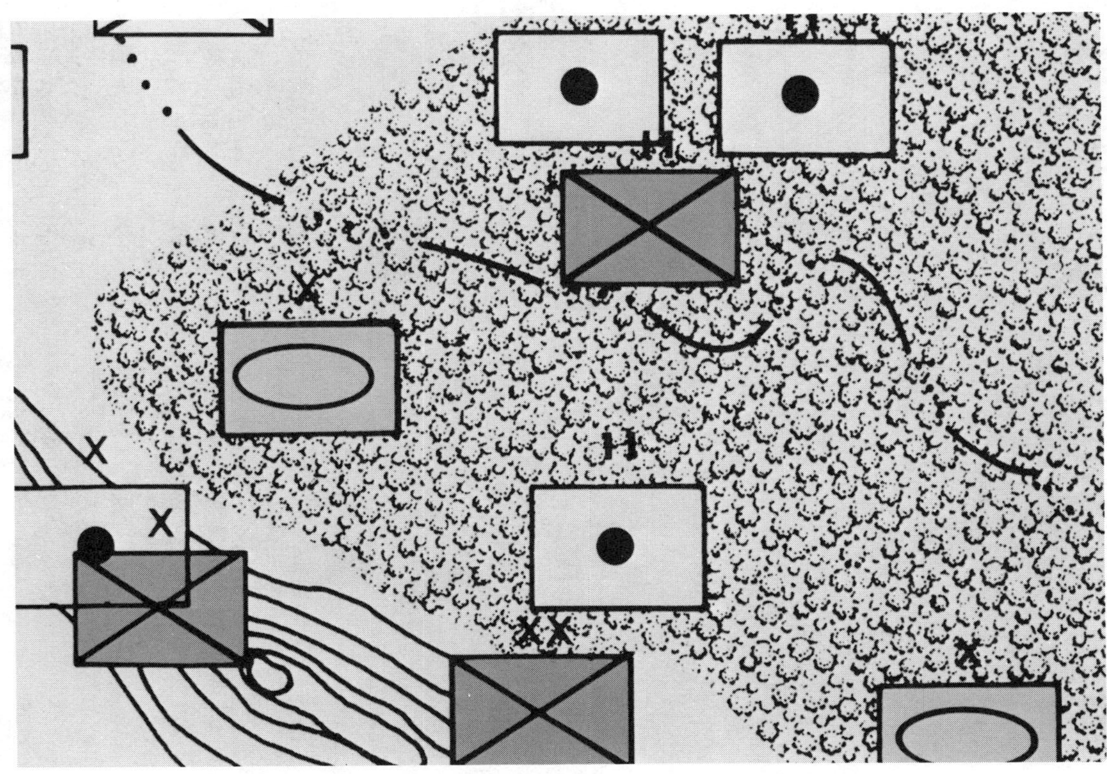

FIGURE 10. Enlarged segment of battlefield situation map with 25% overlap of Figure 9

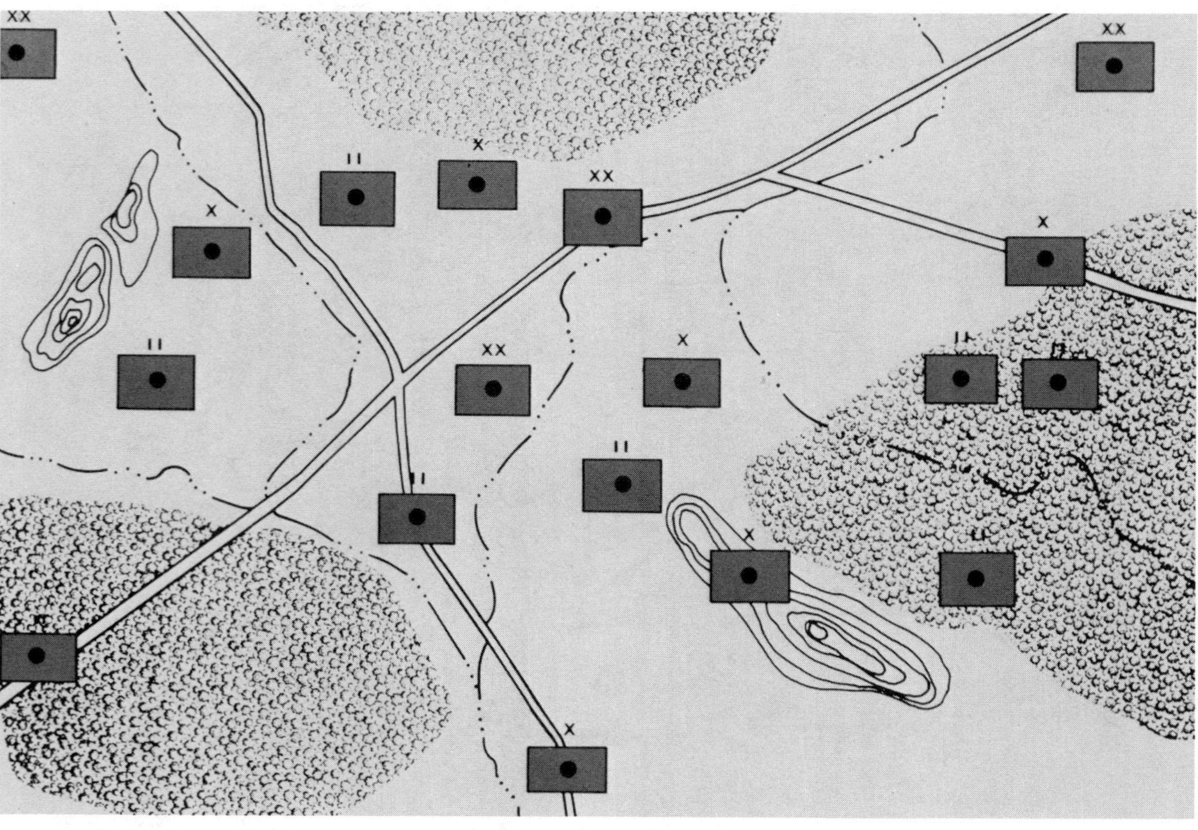

FIGURE 11. Selective call-up of artillery units

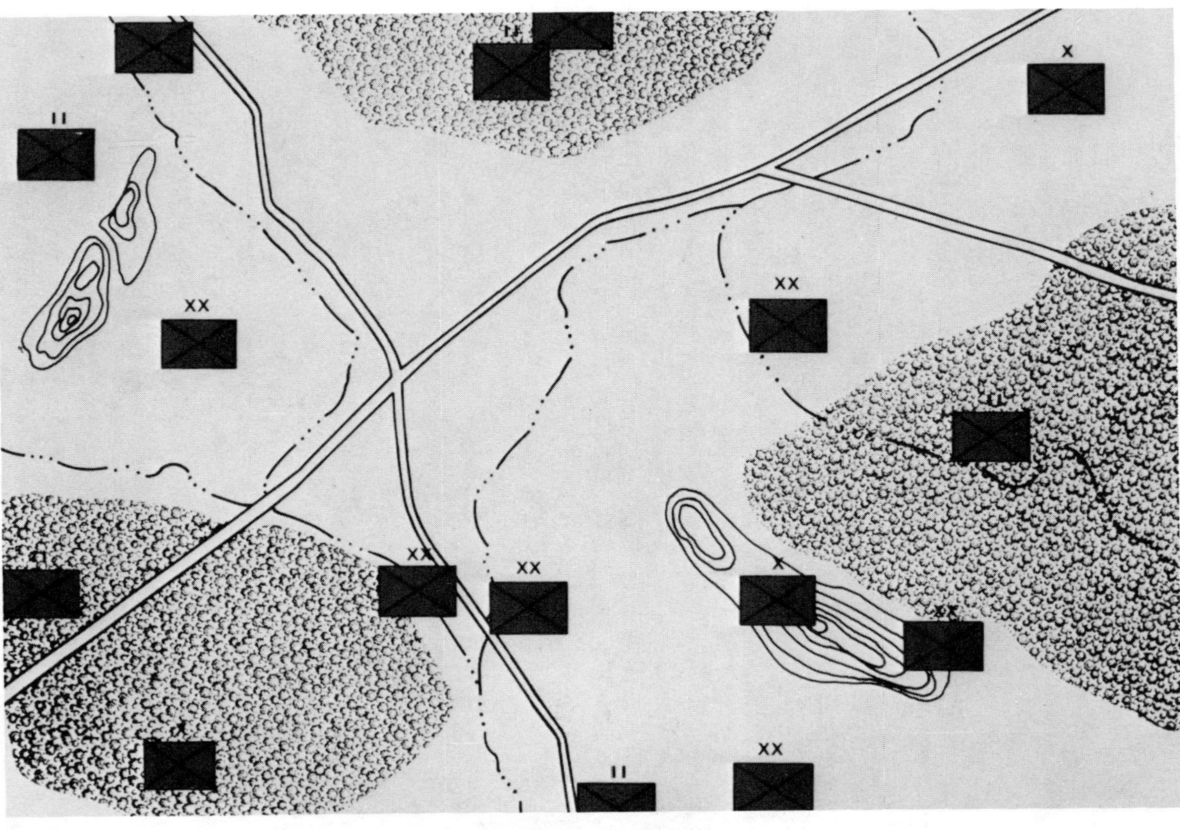

FIGURE 12. Selective call-up of infantry units

211

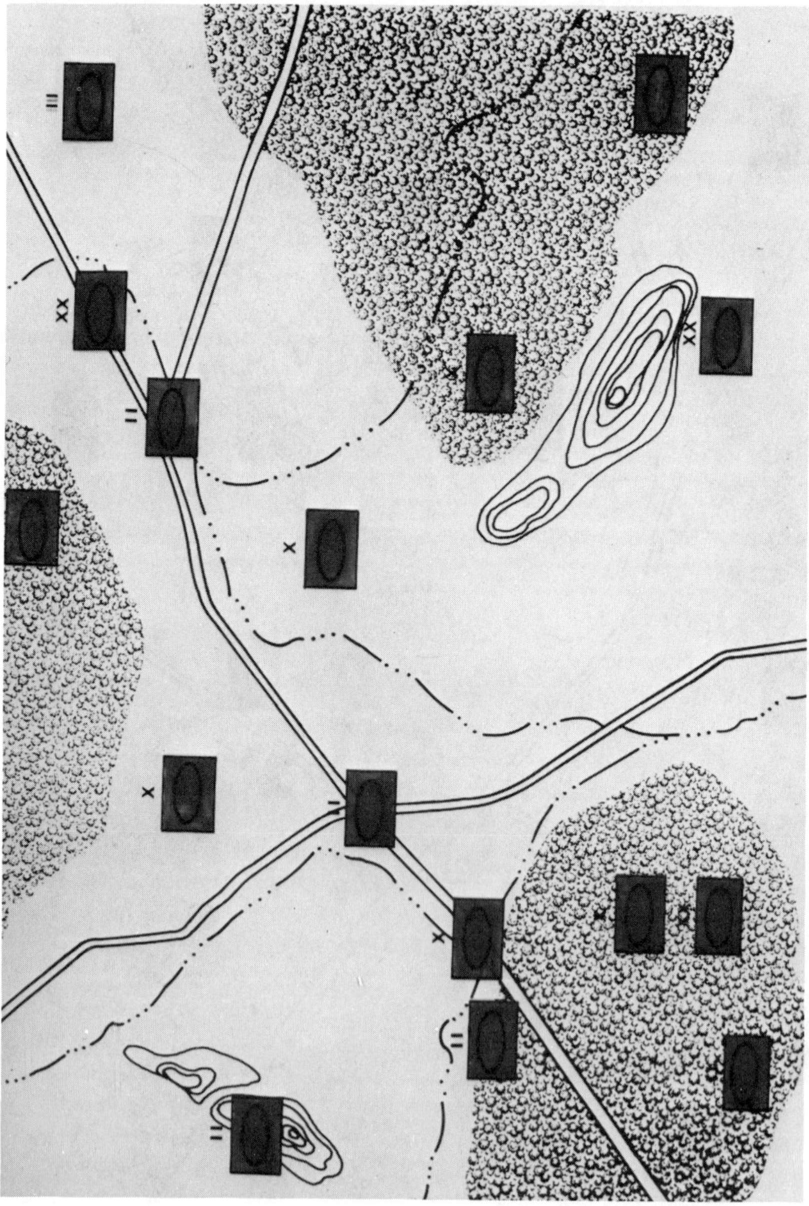

FIGURE 13. Selective call-up of armor units end

2. Double coding dramatically increases user performance in information extraction. A redundant code should augment symbolic information, rather than add new symbols.

3. Color is an extremely powerful coding technique for segregating information, but use it to assist the user at the primary or first level of interest. If a display is to be used for multiple tasks, or if the system designer does not know what information is of prime importance to the user, color coding should be placed under user control.

4. Techniques for reducing symbol density by segmenting displayed information should be standard functions on automated systems. Topographic information can be displayed sequentially, and an overlap between views facilitates user integration of information.

5. Selective call-up seems to be an attractive solution to high-density, cluttered displays. However, penalties are incurred in the amount of time required to "build" a particular information display. Research is needed to examine means of effectively using this option.

DISCUSSION

Coding and sequencing of information have been shown to effectively aid users in extracting desired information from complex displays. Clearly, none of the recommendations made here will provide an unconditional solution to graphic communication problems. However, a more rigorous, empirically based set of guidelines is possible for coding and sequencing use if the findings and recommendations given above are pooled with concurrent and emerging findings in related applied areas. On the topic of color coding, for example, consider the value of work (Tullis, 1981) that derived conclusions about coding and structuring of formats for a computer-based system designed to diagnose problems in telephone lines. Similarly, consider Christ and Corso's (1975) effort that compared various coding schemes (letters, digits, shapes versus color) in a number of simulated air traffic control tasks to investigate the efficacy of a color code to reduce the complexity of this type of task. Findings in these areas are most applicable to the specific context, although certain general principles can emerge: color helps user acceptance of systems; performance may be enhanced equally by color or achromatic codes.

Finally, when data is not available to solve a particular formatting problem, Gellman (1979) suggested the use of a "quick-and-dirty" experiment. It takes no more than a couple of hours, and can provide some predictive measures of the display's usefulness. First, proceed by using preference judgments to choose both the best and the worst formats for the same screen. Second, secure two volunteers to perform a display-related task and then take three or four measures of the performance time with each display format. Finally, do a cost/benefit analysis of the formatting techniques based upon the cost of implementing each technique versus the user's speed of task performance. For example, if redundant color coding only provided the user with an additional second over a redundant non-color code, then the hardware and software expenditure to implement color may be unjustifiable. Based upon a small research investment, you have some indication that color is not necessary; initial preference judgments typically would lead to the opposite conclusion. Of course, this simplistic form of experimentation provides only one more tool along with more rigorous research and available data for choosing display formats.

CONCLUSION

An integrated program of guideline development for display coding and sequencing consists of applied, task-specific studies followed by various comparative and collative analyses. From this approach, specific questions are anwered and more general principles are derived. This use of research and synthesis is well suited to the evaluation of standards for display format techniques.

REFERENCES

Bloomfield, J. R., Beckwith, W. E., Emerick, J., Marmurek, H. H., Tei, B.E., & Traub, B. H. *Visual search with embedded targets* (US Army Research Institute. Tech. Rep. TR 78-TH8), Alexandria, VA, December 1978.

Christ, R. E. Review and analysis of color coding research for visual displays. *Human Factors,* 1975, *17*(6), 542–570.

Christ, R. E., & Corso, G. M. *Color research for visual displays final report* (Office of Naval Research CR-213-102-3), Alexandria, VA, July 1975.

Ciccone, D. S., Samet, M. G., & Channon, J. B. *A framework for the development of improved tactical symbology* (US Army Research Institute Tech. Rep. No.403), Alexandria, VA, August 1979.

Engel, S. E., & Granda, R. E. *Guidelines for man/display interfaces,* (TR 00.2720). Poughkeepsie, NY: IBM, Poughkeepsie Laboratory, 1975.

Gellman, L. H. *Display formatting techniques and applications.* Paper presented at Army 2nd Computer Graphics Workshop, Virginia Beach, 1979.

Granda, T. M. A comparison between a standard map and a reduced detail map with a simulated tactical operations system (SIMTOS). (US Army Research Institute Technical Paper 274), Alexandria, VA, June 1976.

Granda, T. M. *An evaluation of visual search behavior on a cathode ray tube utilizing the window technique* (US Army Research Institute Technical Paper 283), Alexandria, VA, February 1978.

Hemingway, P. W., Kubala, A. L., & Chastain, G. D. *Study of symbology for automated graphic displays* (US Army Research Institute Tech. Rep. 79-A18), Alexandria, VA, May 1979.

Hochberg, J. In the mind's eye. In R. N. Haber (Ed.), *Contemporary theory and research in visual perception.* New York: Holt, Rinehart and Winston, 1968.

Koffka, K. *Principles of gestalt psychology.* New York: Harcourt, Brace, 1935.

Krebs, M. J., Wolf, J. D., & Sandvig, J. H. *Color display design guide, prepared for Office of Naval Research*. (Office of Naval Research Report No. CR213-1382f). October 1978.

Landee, B. M., Samet, M. G., & Foley, D. R. *A task based analysis of information requirements of tactical maps* (US Army Research Institute Tech. Rep. No. 397), Alexandria, VA, August 1979.

Landee, B. M., Samet, M. G., & Gellman, L. H. *A methodology for elicitation and organization of tactical information requirements* (US Army Research Institute Tech. Rep.), Alexandria, VA, In preparation, 1980.

Meister, D., & Sullivan, D. J. *Guide to human engineering design for visual displays*. Canoga Park, CA: Defense Systems Division, Bunker-Ramo Corporation, Alexandria, VA, August 1969.

Moses, F. L., & Maisano, R. E. *User performance under several automated approaches to changing displayed maps* (US Army Research Institute Technical Paper 366), Alexandria, VA, June 1979.

Sidorsky, R. C. Color coding in tactical displays: Help or hindrance? (US Army Research Institute Technical Paper), Alexandria, VA. In preparation, 1980.

Swets, J. A. *Human monitoring behavior in a command/control/communication system*. Prepared for Advanced Research Projects Agency, Contract No. MDA-903-76-C-0207, October 1976.

Synder, H. C., & Maddox M. E. *Information transfer from computer generated Dot Matrix Displays*, US Army Research Office (ARO 78-1), Alexandria, VA, October 1978.

Tullis, T. S. An evaluation of alphanumeric, graphic, and color information displays. *Human Factors*, 1981, *23*(5), 541–550.

Vicino, F. L., Andrews, R. S., & Ringel, S. Conspicuity coding of updated symbolic information (US Army Research Institute Technical Research Note 152), Alexandria, VA, May 1965.

Author Index

Goudie, G., 166, *178*
Gould, J. D., 4, 6, 8, 12, 13, 18, *22, 25,* 96,
    109, 180, *193*
Graham, J. W., 56, *78*
Granda, R. E., 195, *214*
Granda, T. M., 199, 206, *214*
Gray, A. H., 156, *178*
Green, T. R. G., 8, *22, 25*
Greeno, J. G., 9, *22*
Gregg, L. W., 12, *22*
Grimes, J. D., 6, *23*
Grossberg, M., 11, *23*
Grosz, B. J., 8, *23*
Guest, D. J., 8, *25*
Guilford, J. P., 128, *135*

H

Hahn, K. W., 57, *78*
Halpern, M., 9, *23*
Hammond, N. V., 5, *20*
Hartley, R., 11, *23*
Hartman, F. R., 15, 18, *23*
Hartwell, S., 5, *23,* 101, 102, 104, *109*
Hayes, J. R., 3, *23*
Heaps, H. S., 57, *78*
Hemingway, P. W., 199, *214*
Hepler, S. P., 180, *193*
Hirsch, R. S., 14, *23*
Hirsh-Pasek, K., 112, 130, *135*
Hochberg, J., 199, *214*
Hodge, M. H., 113, *135*
Hsia, H. J., 15, *23*
Huckle, B. A., 146, *148*
Hunnicut, S., 15, *25*

J

Jackson, M. D., 95, 107, *109*
Jagacinski, R. J., 15, *21*
Jende, M. S., *148*
Jensen, K., *53*
Johannsen, G., 18, *23*
Johnsen, J. E., 172, *178r*
Just, M. A., 3, 23

K

Kahneman, D., 14, *23*
Kanter, J., 10, *23*

Kaplan, S. J., 81, *91*
Keenan, S. A., 13, *23*
Kelly, M. J., 6, *23,* 83, 84, *91,* 172, 173, 176,
    178
King, J., 3, *21*
Kintsch, W., 8, *23*
Knapp, B. G., 14, *23*
Knuth, D., 44, *53*
Koether, M. E., 14, *21*
Koffka, K., 199, *214*
Konz, S., 180, *193*
Krebs, M. J., 200, *215*
Kubala, A. L., 199, *214*
Kucera, H., 173, *178*
Kurzweil, R., 166, 168, *178*

L

Ladefoged, P., 151, *178*
Lambert, J. V., 113, *135*
Landauer, T. K., 5, 7, *22, 23,* 28, 101, 102,
    104, 107, *108, 109*
Landee, B. M., 199, *215*
Lane, N. E., 113, *135*
Larkin, J., 51, *53*
Ledgard, H. F., 5, 11, *23,* 104, 106n, 107, *109*
Lesk, M. E., 97, *109*
Limb, J. O., 14, 15, *23*
Lin, K., 166, *178*
Lindsay, P., 2, *23*
Lindsay, R. K., 172, *178*
Lochhead, J., 48, *54*
Long, J. B., 5, *20*
Love, T., 52, *53*

M

Macdonald, N. H., 13, *23*
MacDonald-Ross, M., 14, *23*
Maddox, M. E., 195, *215*
Maisano, R. E., 6, 18, *22,* 206, *215*
Malhotra, A., 6, 7, 10, *21, 23, 24,* 84, *91*
Markel, J. D., 156, *178*
Marmurek, H. H., 197, *214*
Marshall, C. F., 6, 18, *22*
Matula, R. E., *24*
Maxemchuk, N. F., 13, *24*
Maxwell, W. L., 56, *77*
Mayer, R. E., 3, 8, 9, *24, 25,* 52, *53, 148*
McBride, D. K., 113, *135*

Subject Index